Lecture Notes in Bioinformatics 5676

Edited by S. Istrail, P. Pevzner, and M. Waterman

Subseries of Lecture Notes in Computer Science

Katia S. Guimarães Anna Panchenko
Teresa M. Przytycka (Eds.)

Advances in Bioinformatics and Computational Biology

4th Brazilian Symposium on Bioinformatics, BSB 2009
Porto Alegre, Brazil, July 29-31, 2009
Proceedings

Series Editors

Sorin Istrail, Brown University, Providence, RI, USA
Pavel Pevzner, University of California, San Diego, CA, USA
Michael Waterman, University of Southern California, Los Angeles, CA, USA

Volume Editors

Katia S. Guimarães
Federal University of Pernambuco, Center of Informatics
Av. Prof. Luiz Freire, s/n, Cidade Universitária, Recife, PE 50740-540, Brazil
E-mail: katiag@cin.ufpe.br

Anna Panchenko
Teresa M. Przytycka
National Center for Biotechnology Information
National Library of Medicine, National Institutes of Health
8600 Rockville Pike, Building 38A, Bethesda, MD 20894, USA
E-mail: {panch, przytyck}@ncbi.nlm.nih.gov

Library of Congress Control Number: 2009930955

CR Subject Classification (1998): H.2.8, F.2.1, I.2, G.2.2, J.3, E.1

LNCS Sublibrary: SL 8 – Bioinformatics

ISSN 0302-9743
ISBN-10 3-642-03222-2 Springer Berlin Heidelberg New York
ISBN-13 978-3-642-03222-6 Springer Berlin Heidelberg New York

springer.com

Printed in Germany

Typesetting: Camera-ready by author, data conversion by Scientific Publishing Services, Chennai, India
Printed on acid-free paper SPIN: 12724642 06/3180 5 4 3 2 1 0

Preface

This volume contains the papers selected for presentation at the 4th Brazilian Symposium on Bioinformatics, BSB 2009, which was held in Porto Alegre, Brazil, during August 29–31, 2009. The BSB symposium had its origins in the Brazilian Workshop on Bioinformatics (WOB). WOB had three editions, in 2002 (Gramado, RS), in 2003 (Macaé, RJ), and in 2004 (Brasília, DF). The change in the designation from workshop to symposium reflects the increase in the quality of the contributions and also in the interest of the scientific community for the meeting. The previous editions of BSB took place in São Leopoldo, RS, in 2005, in Angra dos Reis, RJ, in 2007, and in Santo André, SP, in 2008.

As evidence of the internationalization of the event, BSB 2009 had 55 submissions from seven countries. Of the 55 papers submitted, 36 were full papers, with up to 12 pages each, and 19 were extended abstracts, with up to 4 pages each. The articles submitted were carefully reviewed and selected by an international Program Committee, comprising three chairs and 45 members from around the world, with the help of 21 additional reviewers. The Program Committee Chairs are very thankful to the authors of all submitted papers, and especially to the Program Committee members and the additional reviewers, who helped select the 12 full papers and the six extended abstracts that make up this book.

The editors would also like to thank the six keynote speakers of BSB 2009: Emil Alexov (Clemson University, USA), Dario Grattapaglia (EMBRAPA, Brazil), Arthur Gruber (USP, Brazil), Anna Panchenko (NIH, USA), Teresa M. Przytycka (NIH, USA), and Eytan Ruppin (Tel Aviv University, Israel).

The conference was sponsored by the Brazilian Computer Society (SBC), and counted on the support of Brazilian sponsoring agencies CNPq and CAPES. Special thanks go to Springer, for agreeing to publish this proceedings volume, and to the Steering Committee of BSB 2009. Last but not least, we would like to express our gratitude to the General and Local Organizing Chairs, Osmar Norberto de Souza and Duncan D.D. Ruiz, and the Local Organizing Committee, whose excellent work and dedication made BSB 2009 an enjoyable experience.

July 2009

Katia S. Guimarães
Anna Panchenko
Teresa M. Przytycka

Preface

Organization

Program Committee Chairs

Katia S. Guimarães	Federal University of Pernambuco (UFPE), Brazil
Anna Panchenko	National Institutes of Health (NIH), USA
Teresa M. Przytycka	National Institutes of Health (NIH), USA

Program Committee

Emil Alexov	Clemson University, USA
Nalvo F. Almeida Jr.	Federal University of Mato Grosso do Sul (UFMS), Brazil
Ana Lucia C. Bazzan	Federal University of Rio Grande do Sul (UFRGS), Brazil
Ana Maria Benko-Iseppon	Federal University of Pernambuco (UFPE), Brazil
Igor Berezovsky	University of Bergen, Norway
Ulisses Braga-Neto	Texas A&M University, USA
Marcelo M. Brigido	National University of Brasília (UNB), Brazil
Helaine Carrer	University of São Paulo (USP) /Piracicaba, Brazil
Andre C.P.L.F. Carvalho	University of São Paulo (USP) /São Carlos, Brazil
Saikat Chakrabarti	National Institutes of Health (NIH), USA
Alberto M.R. Dávila	Oswaldo Cruz Institute, Brazil
Frank Dehne	Carleton University, Canada
Zanoni Dias	State University of Campinas (UNICAMP), Brazil
Carlos E. Ferreira	University of São Paulo (USP) /São Paulo, Brazil
Jessica Fong	National Institutes of Health (NIH), USA
Oxana Galzitskaya	Institute of Protein Science, Russia
Ronaldo F. Hashimoto	University of São Paulo (USP) /São Paulo, Brazil
Maricel Kann	University of Maryland, Baltimore County (UMBC), USA
Carl Kingsford	University of Maryland, College Park (UMD), USA
Eugene Krissinel	European Bioinformatics Institute (EBI), UK
Ney Lemke	São Paulo State University (UNESP), Brazil
Thomas Madej	National Institutes of Health (NIH), USA
Wojtek Makalowski	University of Münster, Germany
Ion Mandoiu	University of Connecticut, USA
Natalia F. Martins	EMBRAPA, Brazil
Wellington S. Martins	Federal University of Goiás (UFG), Brazil
Jose Carlos M. Mombach	Federal University of Santa Maria (UFSM), Brazil
Osmar Norberto-de-Souza	Pontifical Catholic University of RS (PUCRS), Brazil
Arlindo Oliveira	INESC, Portugal
Carlos A. B. Pereira	University of São Paulo (USP) /São Paulo, Brazil

Ron Y. Pinter	Technion, Israel
Carlos H.I. Ramos	UNICAMP, Brazil
Cenk Sahinalp	Simon Fraser University, Canada
Francisco M. Salzano	Federal University of Rio Grande do Sul (UFRGS), Brazil
João Carlos Setubal	Virginia Bioinformatics Institute (VBI), USA
Roded Sharan	Tel Aviv University, Israel
Hagit Shatkay	Queen's University, Canada
Benjamin Shoemaker	National Institutes of Health (NIH), USA
Mona Singh	Princeton University, USA
Marcílio M.C.P. de Souto	Federal University of Rio Grande do Norte (UFRN), Brazil
Jerzy Tiuryn	University of Warsaw, Poland
Ana Teresa Vasconcelos	National Lab. of Scientific Computation (LNCC), Brazil
Maria Emília M.T. Walter	National University of Brasília (UNB), Brazil
Alex Zelikovsky	Georgia State University, USA
Jie Zheng	National Institutes of Health (NIH), USA

Additional Reviewers

Gulsah Altun	Scripps Res. Institute, USA
Irina Astrovskaya	Georgia State University, USA
Ananth Bommakanti	University of Maryland, Baltimore County (UMBC), USA
Janusz Dutkowski	University of Warsaw, Poland
Katti Faceli	Federal University of São Carlos (UFSCar), Brazil
André Fujita	University of Tokyo, Japan
Anna Gambin	University of Warsaw, Poland
Yang Huang	National Institutes of Health (NIH), USA
Andre Kashiwabara	University of São Paulo (USP) /São Paulo, Brazil
Attila Kertesz-Farkas	University of Maryland, Baltimore County (UMBC), USA
Boguslaw Kluge	University of Warsaw, Poland
Giovani R. Librelotto	Federal University of Santa Maria (UFSM), Brazil
Fabrício Lopes	Federal University of Technology, Paraná (UTFPR), Brazil
Carlos Maciel	University of São Paulo (USP) /São Paulo, Brazil
Sílvio B. de Melo	Federal University of Pernambuco (UFPE), Brazil
Dilvan de A. Moreira	University of São Paulo (USP) /São Paulo, Brazil
Carlos Santos	University of São Paulo (USP) /São Paulo, Brazil
José Soares	University of São Paulo (USP) /São Paulo, Brazil
Rajesh Thangudu	National Institutes of Health (NIH), USA
Manoj Tyagi	National Institutes of Health (NIH), USA
Nir Yosef	Tel Aviv University, Israel

General Chairs and Local Organizing Chairs

Osmar Norberto de Souza	Pontifical Catholic University of RS (PUCRS)
Duncan D.D. Ruiz	Pontifical Catholic University of RS (PUCRS)

Local Organizing Committee

Ana Winck	Pontifical Catholic University of RS (PUCRS)
Carla Aguiar	Pontifical Catholic University of RS (PUCRS)
Christian Quevedo	Pontifical Catholic University of RS (PUCRS)
Danieli Forgiarini	Pontifical Catholic University of RS (PUCRS)
Elisa Cohen	Pontifical Catholic University of RS (PUCRS)
Karina Machado	Pontifical Catholic University of RS (PUCRS)

Steering Committee

Maria Emília M.T. Walter (Chair)	National University of Brasília (UNB)
Ana L.C. Bazzan	Federal University of Rio Grande do Sul (UFRGS)
André C.P.L.F. de Carvalho	University of São Paulo (USP) /São Carlos
Katia S. Guimarães	Federal University of Pernambuco (UFPE)
Osmar Norberto-de-Souza	Pontifical Catholic University of RS (PUCRS)
Francisco M. Salzano	Federal University of Rio Grande do Sul (UFRGS)
João Carlos Setubal	Virginia Bioinformatics Institute (VBI, USA)

Table of Contents

Algorithmic Approaches for Molecular Biology Problems

Genotype Tagging with Limited Overfitting 1
Irina Astrovskaya and Alex Zelikovsky

Constraint Programming Models for Transposition Distance Problem ... 13
Ulisses Dias and Zanoni Dias

Comparison of Spectra in Unsequenced Species 24
Freddy Cliquet, Guillaume Fertin, Irena Rusu, and Dominique Tessier

Micro-array Analysis

BiHEA: A Hybrid Evolutionary Approach for Microarray Biclustering 36
Cristian Andrés Gallo, Jessica Andrea Carballido, and Ignacio Ponzoni

Using Supervised Complexity Measures in the Analysis of Cancer Gene Expression Data Sets 48
Ivan G. Costa, Ana C. Lorena, Liciana R.M.P. y Peres, and Marcilio C.P. de Souto

Quantitative Improvements in cDNA Microarray Spot Segmentation.... 60
Mónica G. Larese and Juan Carlos Gómez

Machine Learning Methods for Classification

SOM-PORTRAIT: Identifying Non-coding RNAs Using Self-Organizing Maps 73
Tulio C. Silva, Pedro A. Berger, Roberto T. Arrial, Roberto C. Togawa, Marcelo M. Brigido, and Maria Emilia M.T. Walter

Automatic Classification of Enzyme Family in Protein Annotation 86
Cássia T. dos Santos, Ana L.C. Bazzan, and Ney Lemke

Representations for Evolutionary Algorithms Applied to Protein Structure Prediction Problem Using HP Model 97
Paulo H.R. Gabriel and Alexandre C.B. Delbem

Comparing Methods for Multilabel Classification of Proteins Using Machine Learning Techniques 109
Ricardo Cerri, Renato R.O. da Silva, and André C.P.L.F. de Carvalho

Comparative Study of Classification Algorithms Using Molecular Descriptors in Toxicological DataBases 121
Max Pereira, Vítor Santos Costa, Rui Camacho, Nuno A. Fonseca, Carlos Simões, and Rui M.M. Brito

In Silico Simulation

Influence of Antigenic Mutations in Time Evolution of the Immune Memory – A Dynamic Modeling 133
Alexandre de Castro, Carlos Frederico Fronza, Poliana Fernanda Giachetto, and Domingos Alves

Short Papers

FReDD: Supporting Mining Strategies through a Flexible-Receptor Docking Database 143
Ana T. Winck, Karina S. Machado, Osmar Norberto-de-Souza, and Duncan D.D. Ruiz

A Wide Antimicrobial Peptides Search Method Using Fuzzy Modeling 147
Fabiano C. Fernandes, William F. Porto, and Octavio L. Franco

Identification of Proteins from Tuberculin Purified Protein Derivative (PPD) with Potential for TB Diagnosis Using Bioinformatics Analysis 151
Sibele Borsuk, Fabiana Kommling Seixas, Daniela Fernandes Ramos, Caroline Rizzi, and Odir Antonio Dellagostin

Mapping HIV-1 Subtype C gp120 Epitopes Using a Bioinformatic Approach 156
Dennis Maletich Junqueira, Rúbia Marília de Medeiros, Sabrina Esteves de Matos Almeida, Vanessa Rodrigues Paixão-Cortez, Paulo Michel Roehe, and Fernando Rosado Spilki

MHC: Peptide Analysis: Implications on the Immunogenicity of Hantaviruses' N protein 160
Maurício Menegatti Rigo, Dinler Amaral Antunes, Gustavo Fioravanti Vieira, and José Artur Bogo Chies

An Ontology to Integrate Transcriptomics and Interatomics Data Involved in Gene Pathways of Genome Stability 164
Giovani Rubert Librelotto, José Carlos Mombach, Marialva Sinigaglia, Éder Simão, Heleno Borges Cabral, and Mauro A.A. Castro

Author Index 169

Genotype Tagging with Limited Overfitting

Irina Astrovskaya* and Alex Zelikovsky

Department of Computer Science, Georgia State University, Atlanta, GA 30303
{iraa,alexz}@cs.gsu.edu

Abstract. Due to the high genotyping cost and large data volume in genome-wide association studies data, it is desirable to find a small subset of SNPs, referred as tag SNPs, that covers the genetic variation of the entire data. To represent genetic variation of an untagged SNP, the existing tagging methods use either a single tag SNP (e.g., Tagger, IdSelect), or several tag SNPs (e.g., MLR, STAMPA). When multiple tags are used to explain variation of a single SNP then usually less tags are needed but overfitting is higher.

This paper explores the trade-off between the number of tags and overfitting and considers the problem of finding a minimum number of tags when at most two tags can represent variation of an untagged SNP. We show that this problem is hard to approximate and propose an efficient heuristic, referred as 2LR. Our experimental results show that 2LR tagging is between Tagger and MLR in the number of tags and in overfitting. Indeed, 2LR uses slightly more tags than MLR but the overfitting measured with 2-fold cross validations is practically the same as for Tagger. 2LR-tagging better tolerates missing data than Tagger.

Keywords: genotype tagging, linear programming, minimum dominating set, hypergraph.

1 Introduction

DNA variations, primarily single nucleotide polymorphisms (SNPs), hold much promises as a basis of the genome-wide association studies. Genome-wide studies are challengeable due to the genotyping cost and computational complexity of the analysis. So it is desirable to find a small subset of SNPs, referred as informative (tag) SNPs, that covers the genetic variation of all SNPs. Tagging saves budget since only tag SNPs are genotyped or, alternatively, reduces complexity of the analysis since the size of the data is reduced.

Statistical covering of a genetic variation of one SNP by another is usually defined in terms of the correlation between SNPs. It is widely accepted that a variation of a SNP is (statistically) covered by a variation of another SNP if their correlation r^2 is greater than 0.8.

The problem of tagging is selecting a minimum number of tag SNPs covering all SNP in the given data. Formally, the problem can be formulated as follows.

Minimum Tag Selection Problem (MTS Problem): given a sample of a population S of n genotypes, each consisting of m SNPs, find a minimum size subset of tag SNPs that statistically covers all other SNPs.

* Supported in part by Molecular Basis of Disease Fellowship, Georgia State University.

K.S. Guimarães, A. Panchenko, T.M. Przytycka (Eds.): BSB 2009, LNBI 5676, pp. 1–12, 2009.

In practice, a small percentage of SNPs values are missing, reducing the power of an association. A missing SNP value may occur either if this SNP was not initially genotyped or if the SNP value did not pass quality control threshold after typing. When a tag SNP missed a value, the genetic variation of untagged SNPs can be also lost, possibly collapsing distinct genotypes into one. Still correct genotypes can be recovered if we use several alternative ways to cover genetic variation of untagged SNPs [13]. In our experiments, we also explore behavior of tagging methods in presence of missing SNPs values.

Several approaches were proposed to optimize a number of chosen tags. They can be divided into two groups: block-based and block-free methods. The block-based tagging (e.g., HapBlock [16]) requires a priori information to define "haplotype blocks". A haplotype block is not well-defined, but usually SNPs are considered to be in one haplotype block if they are in strong linkage disequilibrium and close to each other [7]. The block-based methods cover an untagged SNP by a tag from the same haplotype block, i.e., only within local contiguous sequence of SNPs where the diversity of haplotypes is low. The block-free methods pick tags across the entire chromosome. Such methods include Avi-Itzhak's method of entropy maximization [1], Halldorson's graphic-based algorithm [8], MLR [11], [10], IdSelect [4], STAMPA [9], Tagger [2], BNtagger [14] .

To represent genetic variation of an untagged SNP, some tagging methods use a single tag, or multiple tags. The Minimum Tag Selection problem for single-tag methods (e.g., IdSelect [4], Tagger [2]) is equivalent to the Minimum Set Cover problem and can be efficiently approximated within $O(logm)$ [15]. We showed that for the multiple-tag methods (e.g., MLR [10], STAMPA [9]), the Minimum Tag Selection problem seems as hard to approximate as the Red-Blue Set Cover [5]. Although, IdSelect [4] and Tagger [2] find close to the optimum number of tag SNPs, MLR [10] uses significantly less tags than Tagger since linear combination of tag SNPs can cover a genetic variation of an untagged SNP. However, multiple linear regression might suffer from overfitting since multiple unrelated tags can participate in covering of an untagged SNP. Additionally, the more tags is used by MLR to cover variation of an untagged SNP, the lower prediction accuracy is obtained (see the Figure 1) [11]. These observations motivate us to bound a number of tags that are used to explain variation of an untagged SNP.

This paper explores the trade-off between the number of tags and overfitting and considers the problem of finding a minimum number of tags when at most two tags can represent variation of an untagged SNP, referred as M2TSP. We show that this problem is hard to approximate and propose an efficient heuristic, referred as 2LR. The experiments show that 2LR is between Tagger and MLR in the number of tags and in overfitting. Indeed, 2LR uses slightly more tags than MLR but the overfitting measured with 2-fold cross validations is practically the same as for Tagger.

Our contributions include:

- 2LR tagging heuristic based on linear regression that uses at most two tags to represent variation of an untagged SNP;
- reduction of Red-Blue Set Cover problem to M2TSP;
- several approaches to solve M2TSP including exact integer linear programming algorithm and greedy heuristics;
- experimental results on HapMap data showing that 2LR tagging

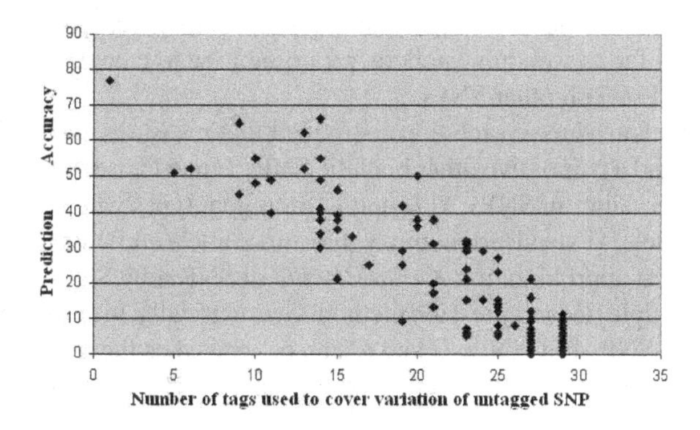

Fig. 1. Correlation between the number of tags being used by MLR to cover variation of an untagged SNP and prediction accuracy [11]

- on overage returns less tags than Tagger without significant increase in overfitting, and
- better tolerates missing data than Tagger.

The reminder of this paper is organized as follows. In Section 2, we introduce correlation for a triple of SNPs and discuss how to adjust the correlation if missing SNP values occur. In the section 3, we formulate the Minimum Tag Selection problem when at most two tag SNPs are allowed to cover variation of an untagged SNP. We also discuss the complexity of the problem. The section 4 gives an exact algorithm based on integer linear programming and its relaxation. In Section 5, we report the results of comparison of three tagging methods - 2LR tagging, MLR and Tagger - on several data sets. Finally, we draw conclusions in Section 6.

2 Correlation for a Triple of SNPs

This section first explains how to measure the correlation between pair of SNPs. Then introduce correlation coefficient for a triple of SNPs. Finally, we discuss how to adjust correlation coefficient to handle missing values.

The correlation coefficient r^2 between two biallelic SNPs A and B measures how genetic drift impacts on linkage disequilibrium [12]. Let $D(A, B)$ denote linkage disequilibrium betweens SNPs A and B and let p and q be frequencies of the major alleles in A and B, respectively. Then correlation between SNPs A, B are given as follows:

$$r^2(A, B) = \frac{D(A, B)^2}{p(1-p)q(1-q)}.$$

The correlation r^2 reaches 1 only if the allele frequencies of A and B are the same and there is an association in haplotypes between the most common alleles at the two SNPs [12].

It would be advantageous to use more than one SNP to cover variation of another SNPs. However, as far as we know, nobody proposed how to calculate correlation between one SNP and several other SNPs.

In this paper, we introduce correlation coefficient that measures correlation between a biallelic SNP A and a pair of two other biallelic SNPs B and C. Let's $\{a_i\}_{i=1}^n$, $\{b_i\}_{i=1}^n$ and $\{c_i\}_{i=1}^n$ are the values in SNPs A, B and C on n genotypes where 1 stands for homozygote major allele, -1 stands for homozygote minor allele and 0 stands for heterozygote in SNP. The best approximation $X = \alpha B + \beta C$ of SNP A by SNPs B and C can be obtained using multiple linear regression method. Geometrically, the best approximation $X = \alpha B + \beta C$ of SNP A by SNPs B and C can be viewed as the projection of vector $\{a_i\}_{i=1}^n$ on the span determined by vectors $\{b_i\}_{i=1}^n$ and $\{c_i\}_{i=1}^n$. (see the Figure 2).

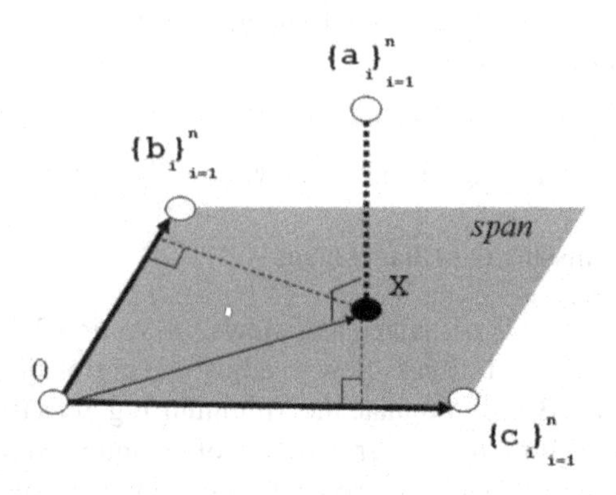

Fig. 2. Geometrically, $X = \alpha B + \beta C$ is projection of $\{a_i\}_{i=1}^n$ on the span determined by $\{b_i\}_{i=1}^n$ and $\{c_i\}_{i=1}^n$

In general, the values of X are real numbers, rather than $\{-1, 0, 1\}$. We round the best approximation $\overline{X}$ as follows. Let $\{\overline{x}_i\}_{i=1}^n$ be the values of rounded best approximation $\overline{X}$ then

$$\overline{x}_i = \begin{cases} 1, if(\alpha b_i + \beta c_i) >= 0.5 \\ -1, if(\alpha b_i + \beta c_i) < -0.5 \\ 0, otherwise \end{cases}, \forall i = \overline{1, n}.$$

A correlation $r^2(A|B, C)$ between SNP A and pair of SNPs B and C is the correlation between SNP A and its rounded best approximation $X = \alpha B + \beta C$ by SNPs B and C:

$$r^2(A|B, C) = r^2(A, X).$$

If the correlation $r^2(A|B, C)$ for a triple of SNPs A, B and C is greater 0.8 then the genetic variation of SNP A is covered by SNPs B and C. In general, the correlation $r^2(A|B, C)$ is not symmetric relation since correlation between SNP A and SNPs B and C does not imply, for example, correlation between SNP B and SNPs A and C, or SNP C and SNPs A and B. Thus, the direction of the correlation $r^2(A|B, C)$ becomes important.

Further, we assume that missing values are distributed randomly and uniformly. Let $\{a_i\}_{i=1}^n$ and $\{b_i\}_{i=1}^n$ be values of SNPs A and B on n genotypes and let $I_{known} \in \overline{1,n}$ be the set of indices of genotypes where values in both SNPs A and B are present. Then we calculate the correlation coefficient as follows:

$$r^2(A,B) = \frac{D(\{a_i\}_{i \in I_{known}}, \{b_i\}_{i \in I_{known}})^2}{p(1-p)q(1-q)},$$

where p and q are frequencies of the major alleles of SNPs A and B on the known data for SNP A and known data for SNP B, respectively. This definition does not introduce additional noise since it is based only on known values.

3 The Minimum Tag Selection Problem

In this section, we first formulate the Minimum Tag Selection problem when a single tag SNP covers a genetic variation of an untagged SNP. Then the problem is generalized to the case when two tag SNPs are allowed to cover variation of an untagged SNP. Finally, the complexity of both problems are discussed.

To represent genetic variation of an untagged SNP, a single-tag tagging methods solve the following problem.

Minimum Tag Selection problem for single-tag covering (M1TS Problem): given a sample of a population S of n genotypes, each consisting of m SNPs, select a minimum size subset of tag SNPs such that genetic variation of each untagged SNP is statistically covered by genetic variation of one tag.

From graph-theoretic point of view, the M1TS problem can be reformulated as the following problem.

Minimum Dominating Set Problem (MDS Problem): given a graph $G = (V, E)$, $E \subseteq V^2$ where each vertex $i \in V$ corresponds to SNP i, $\overline{i = 1, m}$, and edges (i, j) connect highly correlated SNPs i and j (usually, $r^2(i, j) \geq 0.8$); find a set $D \subset V$ of a minimum cardinality such that every vertex $v \in V$ either belongs to D, or is adjacent to vertex $w \in D$.

The Minimum Dominating Set problem is known to be NP-hard, and unless $P = NP$, it can not be approximated better than $O(logm)$ by reducing from Set Cover problem [15].

When two tag SNPs are allowed to cover a variation of an untagged SNP, the M1TSP problem is generalized as follows.

Minimum Tag Selection problem for two-tag covering (M2TS Problem): given a sample of a population S of n genotypes, each consisting of m SNPs, select a minimum size subset of tag SNPs such that genetic variation of each untagged SNP is statistically covered by genetic variation of at most two tags.

From graph-theoretic point of view, the M2TS problem corresponds to Minimum Dominating Set problem on the following hypergraph $H = (V, E)$, $E \subseteq V^2 \bigcup V^3$. Each vertex $i \in V$ corresponds to SNP i, $\overline{i = 1, m}$. There are two types of hyperedges in

H: hyperedges connecting only two vertices (further referred as edges) and hyperedges connecting 3 vertices. As earlier, a bidirectional edge (i, j) connects highly correlated SNPs i and j (usually, $r^2(i, j) \geq 0.8$). A hyperedge (i, j, k) exists if SNP i is highly correlated with SNPs j and k, that is correlation between three SNPs $r^2(i|j, k) \geq 0.8$. To reduce complexity, if at least one of the edges (i, j) or (i, k) exists, we do not add hyperedge (i, j, k) into the hypergraph H. Indeed, high value of the correlation between SNP i and SNPs j and k is due the fact that i is highly correlated with one of the other SNPs, j or k; that is, the hyperedge does not refer to new information. Obviously, since correlation for a triple SNPs $r^2(i|j, k)$ is not symmetric, the existence of the hyperedges (j, i, k) or (k, i, j) does not follow from the hyperedge (i, j, k) although it implies hyperedge (i, k, j).

Minimum Dominating Set problem on a hypergraph $H = (V, E)$: asks for a subset $D \subset V$ of a minimum cardinality such that every vertex $v \in V$ either belongs to D, or is connected by an edge to vertex in D, or is connected by a hyperedge to two vertices in D.

The Minimum Dominating Set problem on Hypergraph is NP-hard, however, an integer linear program gives an exact solution for instances of smaller size. For instances of large sizes, the relaxed linear program returns the approximation of the exact solution (for further details, see Section 4).

The M2TS problem can not be efficiently approximated: it seems as hard to approximate as the following problem.

Red-Blue Set Cover: given two disjoint finite sets, set R of red elements and set B of ble elements, let $S \subseteq 2^{R \bigcup B}$ be a family of subsets of $R \bigcup B$; find a subfamily $C \subseteq S$ that covers all blue elements, but covers the minimum possible number of red elements.

Indeed, if red elements represent possible tag SNPs and blue elements represent untagged SNPs, the M2TS problem is equivalent to Red-Blue Set Cover. If the set of tag candidates and set of SNPs to cover are not intersected than this special case of M2TS problem is the special case of Red-Blue Set Cover where every set contains only one blue and two red elements. It was shown [6], [5] that unless $P = NP$ this special case of the Red-Blue Set Cover can not be approximated to within $O(2^{log^{1-\delta} n})$, where $\delta = \frac{1}{loglog^c n}, \forall c < 1/2$.

4 LP Formulation for M2TS Problem

As follows from the previous section, each instance of M2TS problem can be viewed as an instance of Minimum Dominating Set problem on the corresponding hypergraph. In the section, Boolean Linear Program (BLP) is given for the small instances of the Minimum Dominating Set problem on the hypergraphs. BLP returns exact solution, but becomes slower for large instances. For a larger size problem, we use BLP relaxation that gives approximate solution.

4.1 Boolean Linear Program (BLP)

For a given instance of the Minimum Dominating Set problem on hypergraph $H = (V, E)$ (equivalently, instance of M2TS problem), we formulate Boolean Linear program (BLP) as follows.

We associate boolean variable x_i with each vertex i in hypergraph $H = (V, E)$ (or, equivalently, with each SNP i). BLP sets x_i to 1 if the vertex i is in D (e.g., the corresponding SNP i is chosen as a tag). Otherwise, x_i is set to 0. Let boolean variables $w_{jk} = x_j \cdot x_k$ be indicators for hyperedges $(i, j, k) \in E$. The variable w_{jk} indicates whether the hyperedge (i, j, k) is chosen by BLP, e.g., vertex i is covered by the hyperedge (i, j, k). Indeed, the variable $w_{jk} = 1$ only if both SNPs j and k are chosen by BLP as tags; otherwise, it is 0.

The BLP's objective is to minimize the number of vertices $i \in D$ subject to the following constraints. First, we require every vertex i either be in D, or be connected by edge to a vertex in D, or be connected by hyperedge to two vertices in D. In the other words, we require SNP i be either a tag SNP, or be covered by either a tag SNP or two tag SNPs. Secondly, BLP should set the indicator variables w_{jk} to 1 if and only if BLP sets to 1 both variables x_j and x_k. Indeed, if x_j or x_k is 0 then $w_{jk} \leq 0$ and BLP sets w_{jk} to 0. Finally, BLP should limits feasible solutions to the boolean variables.

Objective

$$\sum_{i=1}^{m} x_i \rightarrow \min$$

Subject to

$$x_i + \sum_{(i,j)\in E} x_j + \sum_{(i,j,k)\in E} w_{jk} \geq 1, i = 1..m$$

$$(x_j + x_k)/2 \leq w_{jk} \leq x_j + x_k - 1, j, k = 1..m$$

$$w_{jk}, x_i \in \{0, 1\}, i, j, k = 1, ..m$$

4.2 BLP Relaxation

BLP gives an exact solution to small instances of the Minimum Dominating Set problem on the hypergraph. For a larger size instances, the BLP relaxation is used.

The BLP relaxation is obtained from the BLP by relaxing the third constraint and allowing fractional solutions to be among feasible solutions, e.g.:

$$w_{jk}, x_i \in [0, 1], i, j, k = 1, ..m.$$

We use several heuristics to round fractional solutions of the BLP relaxation, including two greedy heuristics and randomized heuristics. The first greedy heuristic chooses k variables with the highest weights assigned by BLP relaxation and set them to 1, meaning the corresponding SNPs are chosen as tag SNPs. The value of the parameter k is the smallest integer among all possible such that chosen tag SNPs can cover genetic variation of every untagged SNP in the sample. The alternative greedy approach is iteratively set to 1's the variable(s) x_i with the highest weight assigned at the current run of BLP relaxation. This (these) x_i are explicitly set to 1 in linear program and BLP relaxation is run to obtain new assignments for the rest of variables. The iterative process terminates when all new assignments become zero. Then SNPs that corresponds to all non-zero x_i's, are chosen as tag SNPs. The second heuristic requires slightly more running time than the first heuristic due to several runs of BLP relaxation. The third heuristic is based on randomized approach where each SNP is chosed as a tag SNP with the probability, equalled to the weight given by the BLP relaxation.

The BLP relaxation returns an approximation of the exact solution; however, it runs much faster than the BLP and can be used for larger instances of M2TS problem.

5 Experimental Results

The section describes several data sets on which we run three tagging methods: 2LR method (BLP and BLP relaxation), MLR and Tagger. We report how well these tagging methods compact the data, measured by a number of tag SNPs. Then we describe how to estimate overfitting in the cross-validation tests and report overfitting for these three methods. Finally, we report data compression of each of three tagging methods if a small percentage of SNP values is missing. Direct comparison with STAMPA is complicated since STAMPA predicts untagged SNPs rather than cover their variation.

5.1 Data Sets

We compared 2LR tagging with Tagger [2] and MLR [10] on several publicly available data sets from HapMap . The largest data set has 156 SNPs with minor frequency allele over 0.01%. Each data set has 90 individuals.

Small percentage of missing SNP values was obtained by randomly removing known values from the given data sets.

5.2 Data Compression

In this paper, we explore tradeoff between the number of tags used to cover a variation of an untagged SNP and possible overfit of a tagging method. First, we compare how many tag SNPs each of three tagging methods chooses in the given data sets. The experimental results on HapMap data sets show that 2LR Tagging (BLP and BLP relaxation) is between Tagger and MLR in the number of tags (see Figure 3(a)). The BLP usually selects almost the same number of tags as MLR. Slight difference in the number of tags between BLP and MLR is a cost due to the limited use of multiple regression in BLP versus unlimited multiple regression in MLR. However, for data sets with more than 100 SNPs BLP becomes significantly slower than the others methods (see Figure 3(b)). The BLP relaxation runs as fast as Tagger and on average finds less tags than Tagger but more tags than BLP and MLR due to approximation nature of the relaxation.

5.3 Measuring Overfit via 2-Cross Validation

Further, we compare overfit of the tagging methods in the cross-validation tests.

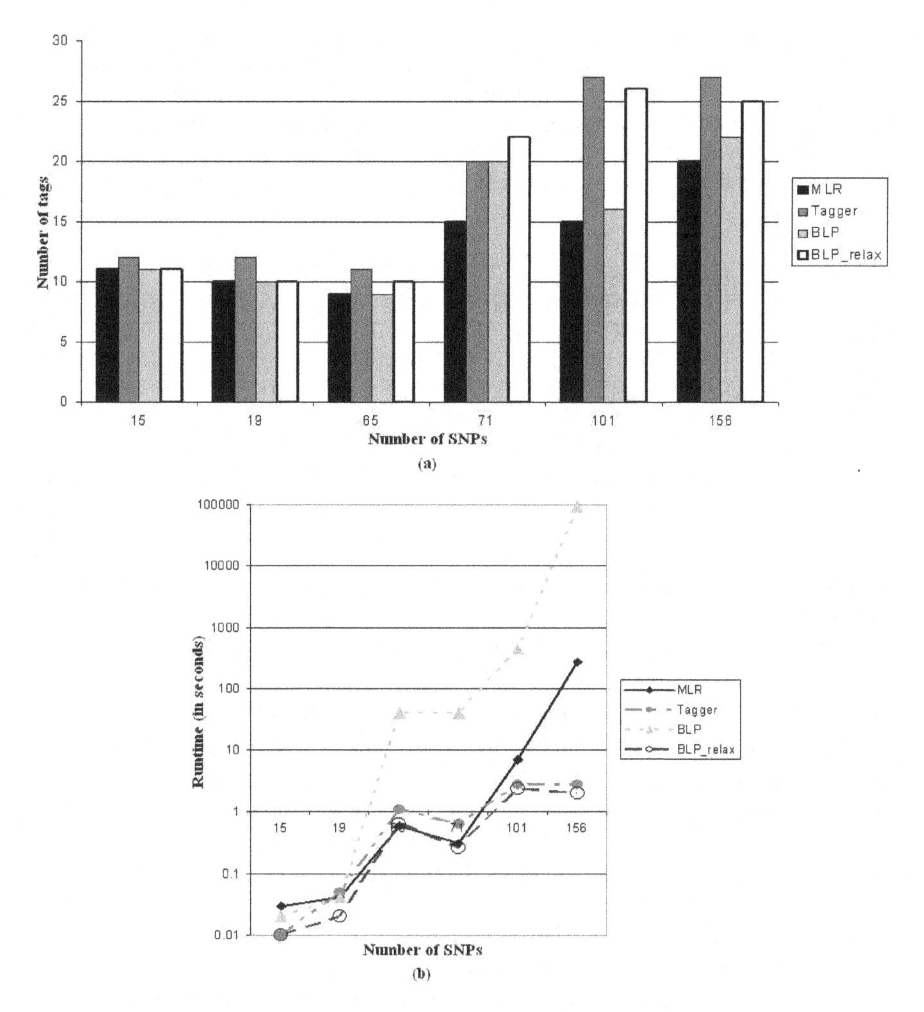

Fig. 3. Performance of MLR, Tagger, 2LR tagging (BLP and BLP relaxation) on HapMap data sets. a) For each tagging method, the number of chosen tags is reported for different number of SNPs in the data sets. b) For each tagging method, its runtime is plotted depending on the number of SNPs in data sets. Runtime is given in seconds.

In 2-fold cross-validation, we partition individuals in the data into two subsets of (approximately) equal size. We use each subset as a test set and the other subset as the corresponding training set. On each training set, we select tag SNPs, then we compute the average covering correlation separately on the training set and its test set if the chosen tags are used. If tagging method suffers from overfitting, we will see significant difference between these two averages. To eliminate dependance between the overfitting value and particular partitioning of the sample, we iteratively run 2-fold cross validations on the heavily permuted sample reasonably large number of times. Then the

Table 1. 2-fold cross validation results for MLR, Tagger and 2LR Tagging (BLP and its relaxation) on ADIPOR2-AA data set with 71 SNPs. The table contains results of several cross-validations. Columns 3-10 show results on different permutations of the data set. For each method, the upper row (NumberTags) gives a number of tags chosen on the training set in particular 2-fold cross-validation run. The lower row give a difference between average covering correlation the on training set and an average covering correlation on the test set, if the chosen tag SNPs are used.

MLR	NumberTags	16	14	15	16	14	13	11	15	14
	ΔR^2	0.0317	0.0422	0.0209	-0.008	0.071	0.0706	0.1843	0.0089	0.0347
Tagger	NumberTags	28	27	29	31	29	30	26	29	27
	ΔR^2	0.0118	0.0437	0.007	-0.0053	0.0258	0.0473	0.1124	0.0205	0.025
BLP	NumberTags	18	17	19	21	18	21	17	18	17
	ΔR^2	0.0489	0.009	0.0349	-0.0135	0.0287	0.0575	0.1343	0.0129	0.012
BLP relaxation	NumberTags	19	19	24	25	22	25	18	21	19
	ΔR^2	0.0114	0.0012	0.0719	0.0289	0.0312	0.0109	0.1334	0.0019	0.04

overfitting of tagging method is reported as the averaged difference in covering correlation among all runs.

Table 1 gives some 2-fold cross validation results for MLR, Tagger, 2LR tagging (BLP and BLP relaxation) on the data set ADIPOR2-AA with 71 SNPs. 2-LR Tagging has less overfit than MLR (on average, 3.6% for 2LR (both BLP and BLP relaxation) vs 5% for MLR), however finds slightly more tags (on average, 19 (21) for BLP (BLP relaxation) and 14 for MLR). On the other hand, 2LR uses less tags than Tagger (19 (or 21 in relaxation) vs. 29 tags on average) while keeping overfit almost at the same rate (3.2% for Tagger and 3.6% for 2LR (both BLP and BLP relaxation)).

Thus, 2LR tagging finds less tag SNPs than Tagger does. However, 2LR tags have practically the same overfit as Tagger tags, meaning smaller set of tags can capture the same amount of a genetic variation as Tagger tags when they being used at the other data sets. The limited used of multiple regression in 2LR tagging rather than unbounded multiple regression in MLR results in slightly larger number of tag SNPs but with smaller overfit to data. In other words, 2LR tagging finds tag SNPs that can capture more genetic variation than MLR tags with a cost of slight increase in the number of tags.

5.4 Missing Data Results

Finally, we explore behavior of three tagging methods on the data set with small percentage of missing data. MLR does not handle missing SNP values and assumes that the data are complete [10]. Thus, we run MLR on the data with imputed the missing SNP values [13]. Our experiments shows that BLP better tolerates missing data than Tagger (see Table 2). The BLP relaxation requires slightly more tags than the BLP to handle missing SNP values. The MLR requires less tag SNPs but the missing values should be imputed first.

Table 2. Comparison of MLR, Tagger and 2LR Tagging (BLP and its relaxation) on ADIPOR2-AA data set with missing SNP values. Number of tags and average correlation is given.

	No Missing Data	1% Missing	2% Missing	3% Missing
Tagger	20	34	48	49
	0.9549	0.9421	0.925	0.9400
MLR	15	15	16	18
	0.984	0.9813	0.989	0.985
BLP	20	24	26	34
	0.945	0.931	0.912	0.8887
BLP relaxation	22	30	45	45
	0.9674	0.9421	0.9235	0.9345

6 Conclusions

In this paper, we explores the trade-off between the number of tags and overfitting and considers the problem of finding a minimum number of tags when at most two tags can represent variation of an untagged SNP, referred as M2TSP. We show that this problem is hard to approximate and propose an efficient heuristic, referred as 2LR. Additionally, we introduce the correlation for a triple of SNPs and show how to adjust correlation to missing SNP values in the data.

The experiments show that 2LR is between Tagger and MLR in the number of tags and in overfitting. Indeed, 2LR uses slightly more tags than MLR but the overfitting measured with 2-fold cross validations is practically the same as for Tagger. 2LR tagging finds smaller subset of tags than Tagger that can capture the same amount of a genetic variation as Tagger's tags when they being used at the other data sets. On the other hand, 2LR tagging finds tag SNPs that can capture more genetic variation than MLR with a cost of slight increase in the number of tags. 2LR tagging also better tolerates missing SNP values than Tagger.

In our future research, we will explore whether overfit of 2LR tagging can be decreased by adding the correlation values between pair or triple of SNPs in BLP and BLP relaxation. Secondly, the correlation between two SNPs does not take into account that 2- SNPs genotypes can be preprocessed by maximum likelihood estimate-based phasing. Finally, since the experiments show that the bounding of tags used for prediction lower the overfit, we will search for the optimum number of tags for predicting an untagged SNP in MLR.

References

1. Avi-Itzhak, H.I., Su, X., De La Vega, F.M.: Selection of minimum subsets of single nucleotide polymorphisms to capture haplotype block diversity. In: Pacific Symposium in Biocomputing, pp. 466–477 (2003)
2. de Bakker, P.I.W., Yelensky, R., Pe'er, I., Gabriel, S.B., Daly, M.J., Altshuler, D.: Efficiency and power in genetic association studies. Nature Genetics 37, 1217–1223 (2005)
3. Calinescu, G.: Private communication

4. Carlson, C.S., Eberle, M.A., Rieder, M.J., Yi, Q., Kruglyak, L., Nickerson, D.A.: Selecting a maximally informative set of single-nucleotide polymorphisms for association analyses using linkage disequilibrium. American Journal of Human Genetics 74(1), 106–120 (2004)
5. Carr, R.D., Doddi, S., Konjevod, G., Marathe, M.V.: On the red-blue set cover problem. In: SODA 2000, pp. 345–353 (2000)
6. Dinur, I., Safra, S.: On the hardness of approximating label cover. ECCC Report 15 (1999)
7. Gabriel, S.B., Schaffner, S.F., Hguyen, H., Moore, J.M., Roy, J., Blumenstiel, B., Higgins, J.: The structure of haplotype blocks in the human genome. Science 296, 2225–2229 (2002)
8. Halldorsson, B.V., Bafna, V., Lippert, R., Schwartz, R., de la Vega, F.M., Clark, A.G., Istrail, S.: Optimal haplotype block-free selection of tagging SNPs for genome-wide association studies. Genome Research 14, 1633–1640 (2004)
9. Halperin, E., Kimmel, G., Shamir, R.: Tag SNP Selection in Genotype Data for Maximizing SNP Prediction Accuracy. Bioinformatics 21, 195–203 (2005)
10. He, J., Zelikovsky, A.: Informative SNP Selection Based on SNP Prediction. IEEE Transactions on NanoBioscience 6(1), 60–67 (2007)
11. He, J., Zelikovsky, A.: Linear Reduction Methods for Tag SNP Selection. In: Proceedings of International Conference of the IEEE Engineering in Medicine and Biology (EMBC 2004), pp. 2840–2843 (2004)
12. Hedrick, P.W., Kumar, S.: Mutation and linkage disequilibrium in human mtDNA. European Journal of Human Genetics 9, 969–972 (2001)
13. Huang, Y.H., Zhang, K., Chen, T., Chao, K.-M.: Approximation algorithms for the selection of robust tag SNPs. In: Jonassen, I., Kim, J. (eds.) WABI 2004. LNCS (LNBI), vol. 3240, pp. 278–289. Springer, Heidelberg (2004)
14. Lee, P.H., Shatkay, H.: BNTagger: improved tagging SNP selection using Bayesian networks. Bioinformatics 22(14), 211–219 (2006)
15. Vazirani, V.V.: Approximation Algorithms. Springer, Heidelberg (2001)
16. Zhang, K., Qin, Z., Chen, T., Liu, J.S., Waterman, M.S., Sun, F.: HapBlock: haplotype block partitioning and tag SNP selection software using a set of dynamic programming algorithms. Bioinformatics 21(1), 131–134 (2005)

Constraint Programming Models for Transposition Distance Problem*

Ulisses Dias and Zanoni Dias

Institute of Computing - Unicamp - Campinas - SP - Brazil
{udias,zanoni}@ic.unicamp.br

Abstract. Genome Rearrangements addresses the problem of finding the minimum number of global operations, such as transpositions, reversals, fusions and fissions that transform a given genome into another. In this paper we deal with transposition events, which are events that change the position of two contiguous block of genes in the same chromosome. The transposition event generates the transposition distance problem, that is to find the minimum number of transposition that transform one genome (or chromosome) into another. Although some tractables instances were found [20, 14], it is not known if an exact polynomial time algorithm exists. Recently, Dias and Souza [9] proposed polynomial-sized Integer Linear Programming (ILP) models for rearrangement distance problems where events are restricted to reversals, transpositions or a combination of both. In this work we devise a slight different approach. We present some Constraint Logic Programming (CLP) models for transposition distance based on known bounds to the problem.

1 Introduction

The problem of inferring evolutive relationship between different organism is one of the most studied problems in biological context. The observable characteristics such as morphological or anatomic traits were the first approaches to classify closely related species.

The genetic data available for many organisms allows more accurate evolutionary inference. The traditional approach uses nucleotide (or amino acid) comparison to find the edit distance, which is the minimum number of local operations, such as insertion, deletion and substitutions that transform a given sequence into another.

The problem with the edit distance approach is that some mutational events affect very large stretches of DNA sequence. For example, a reversal is a genome rearrangement event that breaks the chromosome at two locations and reassembles the piece in the reversed order. When a reversal occur the result is a DNA sequence with essentially the same features, but the edit distance indicates highly diverged genomes.

* This work is partially sponsored by CNPq (472504/2007-0 and 479207/2007-0) and FAPESP (2007/05574-4).

K.S. Guimarães, A. Panchenko, T.M. Przytycka (Eds.): BSB 2009, LNBI 5676, pp. 13–23, 2009.

Genome rearrangement field focus on the comparison of the positions of the same blocks of genes on distinct genomes and on the rearrangement events that possibly transformed an genome into another. Many researches show that rearrangements are common on plant, mammal, virus and bacteria [2,10,15,16,19,23].

In this paper we focus on a rearrangement event called transposition. The transposition exchanges two adjacent blocks of any size in a chromosome. The transposition distance problem is to find the minimum number of transpositions that transform one genome into another.

Although some tractables instances were found [20,14], the transposition distance problem is still open, we do not know any NP-hardness proof and there are no evidences that an exact polynomial algorithm exists. Several approximation algorithms for this problem have been proposed. The best approximation ratio is 1.375 and was presented by Elias and Hartman [12]. The proof of the algorithm was assisted by a computer program. Other approximation algorithms were proposed by Bafna and Pevzner [3], Christie [8], Hartman and Shamir [17], Mira et al. [22], Walter, Dias and Meidanis [24] and Benot-Gagn and Hamel [4].

Recently, Dias and Souza [9] presented an exactly formulation using Integer Linear Programming (ILP) models, following the research line of Caprara, Lancia and Ng [5,6,7]. This work proposes Constraint Logic Programming (CLP) models for transposition distance based on known bounds to the problem. We recommend Marriott and Stuckey [21] as introduction for CLP.

The paper is divided as follows. Section 2 provides the important concepts and definitions used throughout the text, Section 3 presents the CLP models, Section 4 discusses some computational tests and some improvements that could be done as future works.

2 Definitions

We assume that the order of block of genes in a genome is represented by a permutation $\pi = (\pi_1 \; \pi_2 \; \ldots \; \pi_n)$, for $\pi_i \in \mathbb{N}$, $0 < \pi_i \leq n$ and $i \neq j \leftrightarrow \pi_i \neq \pi_j$. A transposition $\rho(i,j,k)$, for $1 \leq i < j < k \leq n+1$, "cuts" an interval $[i, j-1]$ of π and pastes between π_{k-1} and π_k. The result is the permutation $\rho\pi = (\pi_1 \; \ldots \; \pi_{i-1} \; \pi_j \; \ldots \; \pi_{k-1} \; \pi_i \; \ldots \; \pi_{j-1} \; \pi_k \; \ldots \; \pi_n)$.

The transposition distance $d(\pi, \sigma)$ between two permutations π and σ, is the minimum number t of transpositions $\rho_1, \rho_2, \ldots \rho_t$ such that $\pi\rho_1\rho_2 \ldots \rho_t = \sigma$. Let ι be the identity permutation, $\iota = (1 \; 2 \ldots \; n)$, the transposition distance between π and σ equals the transposition distance between $\sigma^{-1}\pi$ and ι. So, without loss of generality, the transposition distance problem is equivalent to the sorting by transpositions problem, which is the transposition distance between a permutation π and the identity, denoted by $d(\pi)$.

A breakpoint for a permutation π is a pair (π_i, π_{i+1}) such that $\pi_{i+1} \neq \pi_i + 1$. We denote by $b(\pi)$ the number of breakpoints in π and $\Delta b(\rho, \pi) = b(\pi\rho) - b(\pi)$ the variation of the number of breakpoints when the transposition ρ is applied on π. Christie [8] introduced the following result about transposition breakpoints.

Lemma 1. *For any permutation π, there is a minimum sequence of transpositions $\rho_1, \rho_2, \ldots, \rho_t$ such that $\pi\rho_1\rho_2\ldots\rho_t = \iota$ and $\Delta b(\rho_i, \pi\rho_1\rho_2\ldots\rho_{i-1}) \leq 0$, for $1 \leq i \leq t$.*

Breakpoints divide a permutation into strips, that are maximal intervals with no breakpoints. Christie [8] shows that every permutation π can be uniquely transformed into a reduced permutation π_{red} with $d(\pi) = d(\pi_{red})$. The transformation consists of removing the first strip if it begins with 1 and last strip if it ends with n, replacing every other strips with its minimal element, and renumbering the resulting sequence in order to obtain a valid permutation.

A transposition acts on three points of a permutation and can decrease the number of breakpoints by at least one and at most three, implying the trivial bounds indicated on Lemma 2 and Lemma 3.

Lemma 2. *For any permutation π, $d(\pi) \geq \frac{b(\pi)}{3}$.*

Lemma 3. *For any permutation π, $d(\pi) \leq b(\pi)$.*

Another useful tool for the transposition distance is the directed edge-colored cycle graph $G(\pi)$ of permutation π proposed by Bafna and Pevzner [3]. The vertex set is defined by first representing each element of π, $|\pi| = n$, as a tuple $(-\pi_i, +\pi_i)$, and after we add the elements 0 and $-(n+1)$ at the left and right extremities respectively. For example, the permutation $\pi = (5\ 2\ 1\ 4\ 3)$ generates the vertices $\{0, -5, 5, -2, 2, -1, 1, -4, 4, -3, 3\}$. The gray edges set is $\{+(i-1), -i\}$, for $1 \leq i \leq n+1$ and the black edges set is $\{(-\pi_i, +\pi_{i-1})\}$, for $1 \leq i \leq n+1$ (Figure 1).

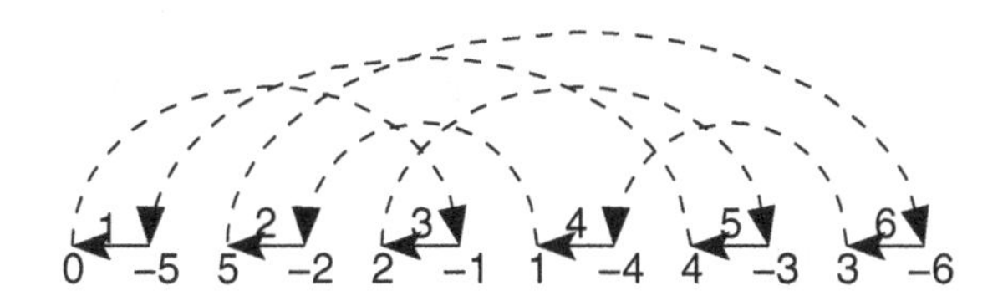

Fig. 1. Cycle graph $G(\pi)$ for $\pi = (5\ 2\ 1\ 4\ 3)$

For each vertex in $G(\pi)$ there is a gray edge paired with a black edge. This implies that there is a unique decomposition of the edges of $G(\pi)$ into cycles of alternating colors. The number of cycles in $G(\pi)$ is denoted by $c(\pi)$.

We denote a cycle with k black edges as a k-cycle. A k-cycle is odd if k is odd, and even otherwise. We denote $c_{odd}(\pi)$ the number of odd cycles on $G(\pi)$. The identity permutation is the only permutation with $n+1$ cycles. The sequence of transpositions that sorts π must increase the number of cycles from $c(\pi)$ to $n+1$.

Let $\Delta c(\rho) = c(\pi\rho) - c(\pi)$ be the variation of the number of cycles and $\Delta c_{odd}(\rho) = c_{odd}(\pi\rho) - c_{odd}(\pi)$ be the variation of the number of odd cycles when the transposition ρ is applied. Bafna and Pevzner [3] showed $\Delta c, \Delta c_{odd} \in \{2, 0, -2\}$, which leads to Lemma 4.

Lemma 4. *For any permutation* π, $d(\pi) \geq \frac{3}{2}(n + 1 - c_{odd}(\pi))$.

Bafna and Pevzner [3] also proved the following upper bound.

Lemma 5. *For any permutation* π, $d(\pi) \leq \frac{3}{4}(n + 1 - c_{odd}(\pi))$.

Another graph called Γ-graph introduced by Labarre [20] can be constructed from a permutation. Given a permutation π, $\Gamma(\pi)$ is the graph with vertex set $\{\pi_1, \ldots, \pi_n\}$ and edge set $\{(\pi_i, \pi_j) | \pi_i = j\}$ (Figure 2).

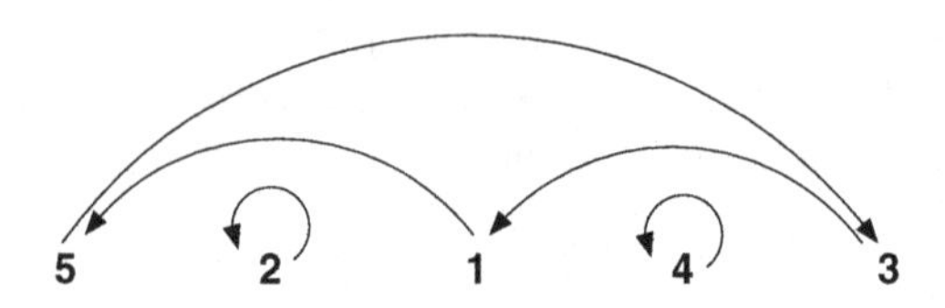

Fig. 2. Γ-graph for $\pi = (5\ 2\ 1\ 4\ 3)$

The Γ-graph allows a unique decomposition of the edges in cycles. Some definitions for Γ-graph are very similar to the cycle graph. For example, given a permutation π, we call a k-cycle in $\Gamma(\pi)$ the cycle with k edges (remember that a k-cycle in $G(\pi)$ is the cycle with k black edges) and we say that a k-cycle in $\Gamma(\pi)$ is odd if k is odd, otherwise it is even. Using this graph Labarre [20] proved the following upper bound.

Lemma 6. *For any permutation* π, $d(\pi) \leq n - c_{odd}(\Gamma(\pi))$.

Let π° be a circular permutation obtained from π by adding the element $\pi_0 = 0$ and let $\pi^\circ_\circ$ be the set formed by all the permutations $\sigma = (m \quad m + \pi_1 \quad m + \pi_2 \ \ldots \ m + \pi_n) \ (mod\ n + 1)$, for $m \in \mathbb{I}$. Eriksson et al. [13] showed that $\sigma \in \pi^\circ_\circ \rightarrow d(\pi) = d(\sigma)$. Labarre [20] used this equivalence [13] to tighten the upper bound on Lemma 6.

Lemma 7. *For any permutation* π, $\pi \neq \iota$, $d(\pi) \leq n - \max_{\sigma \in \pi^\circ_\circ} c_{odd}(\Gamma(\sigma))$.

A second heuristic was obtained through reduction.

Lemma 8. *For any permutation* π, $\pi \neq \iota$, $d(\pi) \leq m - \max_{\sigma \in (\pi_{red})^\circ_\circ} c_{odd}(\Gamma(\sigma))$, *where* $|\pi_{red}| = m$.

3 CLP Models for Sorting by Transposition

In this section we present CLP formulations for the problem of sorting by transposition. We used the lower and upper bounds presented on previous section to generate formulations based on Constraint Satisfaction Problems (CSP) and Constraint Optimization Problems (COP) theory. We define the formulations using prolog-like Marriott's notation [21] as much as possible.

The representation of the permutation (1) and the effects of the transposition (2) can be seen as the same way we described the problem. For instance, a permutation π is a list of elements $(\pi_1, \pi_2, \ldots, \pi_n)$ where $\pi_i \in \mathbb{N}$, $0 < \pi_i \leq n$ and $\pi_i \neq \pi_j$ for $i \neq j$.

$$
\begin{aligned}
& permutation(\pi, N) \text{ :-} \\
& \quad length(\pi, N), \\
& \quad \pi :: [1 \, .. \, N], \\
& \quad all\_different(\pi).
\end{aligned}
\tag{1}
$$

Remember that prolog denotes variables by strings starting with an upper case letter or "_" (the underscore) if the variable is anonymous. We extend the notation to represent the permutations with greek letters π and σ, which are lists. The construction $X :: [A \, .. \, B]$ means that X (or every element in X if it is a list) ranges over the interval $[A..B]$.

A transposition $\rho(i, j, k)$, $0 < i < j < k \leq n$, has the effect of splitting the list in four sub-lists $C_1 C_2 C_3 C_4$ where $C_1 = (\pi_1, \ldots, \pi_{i-1})$, $C_2 = (\pi_i, \ldots, \pi_{j-1})$, $C_3 = (\pi_j, \ldots, \pi_{k-1})$ and $C_4 = (i_k, \ldots, i_n)$, and joining them to form $\pi\rho = C_1 C_3 C_2 C_4$. Notice that C_1 and C_4 could be empty.

$$
\begin{aligned}
& transposition(\pi, \sigma, I, J, K) \text{ :-} \\
& \quad permutation(\pi, N), \\
& \quad permutation(\sigma, N), \\
& \quad 1 \leq I < J < K \leq N, \\
& \quad split(\pi, I, J, K, C_1, C_2, C_3, C_4), \\
& \quad \sigma = C_1, C_3, C_2, C4.
\end{aligned}
\tag{2}
$$

We first model the problem as a CSP, but the number of variables is unknown because we need the distance $d(\pi)$ to set the constraints and the variables that represent the permutations. For this reason, we pick a candidate distance $T \in [L..U]$, where L is a known lower bound and U is a known upper bound, and try to find an appropriate combination of T transpositions. If the CSP with the T candidate fails, we choose another candidate by systematically incrementing the value of T.

It is important to realize that the upper bound U does not matter right now, because we check the value of T in a bottom-up strategy and for definition we will not check for a value higher than any upper bound.

$$
\begin{aligned}
& distance(\iota, 0, \_Model). \\
& distance(\pi, T, Model) \text{ :-} \\
& \quad bound(\pi, Model, LowerBound, UpperBound), \\
& \quad T :: [LowerBound \, .. \, UpperBound], \\
& \quad indomain(T), \\
& \quad transposition(\pi, \sigma, \_I, \_J, \_K), \\
& \quad distance(\sigma, T - 1, Model).
\end{aligned}
\tag{3}
$$

All the CSP models have the above structure, varying only on the bounds we used. In addition, we introduced a model without any bounds for comparison purposes. We call **def_csp**, **br_csp** and **cg_csp** the model that does not use any lower bound, the model that uses the breakpoint lower bound showed in Lemma 2, and the model that uses the cycle graph lower bound showed in Lemma 4, respectively. Assume that the predicate *bound* on (3) receives an atom on variable *Model* representing the models, this atom unifies with the clause that returns the appropriate lower bound.

An important issue on (3) is the use of the anonymous variables $\_I$, $\_J$ and $\_K$. Using anonymous variables we let the solver assign the values based on the constraint $1 \leq \_I < \_J < \_K \leq N$ defined on (2). The solver will assign to the variables the least values possible in the domains and if these values do not lead to an optimal sequence of transposition, then it will be updated based on the same approach.

The predicate $indomain(X)$ on (3) gets the domain of the variable X and chooses the minimum element in it. If a fail backtracks to *indomain*, the element that generated the fail will be removed from the domain and another value will be chosen.

It is possible to improve the CSP models by using the reduced permutations instead of the permutation itself. The reduced permutation has possibly a smaller set, so it is probably easier than the original permutation. To use the reduced permutation we need to create a new predicate *distance_red* to calculate the permutation distance.

$$\begin{aligned}
&distance_red(\pi, 0, _Model) \text{ :- } reduce(\pi, [\,]).\\
&distance_red(\pi, T, Model) \text{ :-}\\
&\quad reduce(\pi, \pi_{red}),\\
&\quad bound(\pi_{red}, Model, LowerBound, UpperBound),\\
&\quad T :: [LowerBound \,..\, UpperBound],\\
&\quad indomain(T),\\
&\quad transposition(\pi_{red}, \sigma, _I, _J, _K),\\
&\quad distance_red(\sigma, T-1, Model).
\end{aligned} \tag{4}$$

Note that we represent the ι_{red} as an empty list, because the operation to reduce a permutation shown on Section 2 eliminates all elements in ι. The models **def_csp**, **br_csp** and **cg_csp** can take advantage of the reduction by creating new models **def_csp_red**, **br_csp_red** and **cg_csp_red**.

Another approach is to model the problem as a COP. This approach needs an upper bound and some modifications on previous predicates. We use the binary variables B to indicate whether a transposition has modified the permutation.

The first predicate that we need to create is the *transposition_cop*. First of all, given a permutation $\rho(i, j, k)$, we add a new clause to allow $(i, j, k) = (0, 0, 0)$. If $(i, j, k) = (0, 0, 0)$ then $\pi\rho = \pi$. We also add a new argument to the *transposition_cop* predicate for the variable B.

$$\begin{aligned}&transposition\_cop(\iota, \iota, 0, 0, 0, 0).\\&transposition\_cop(\pi, \sigma, I, J, K, 1) \text{ :- } transposition(\pi, \sigma, I, J, K).\end{aligned} \tag{5}$$

To accomplish the COP models we implemented the *distance_cop* predicate (6), which set up the variables B using the upper bound and constrains the permutations by making the permutation $\pi_k = \pi_{k-1}\rho_k$. The *length/2* predicate is a prolog built-in used to create a list of non instantiated variables of a given size. The cost function $Cost$ is the sum of the variables B associated with each transposition ρ_k, $Cost = \sum_{k=1}^{UB} B_k$, for UB being a known upper bound. The distance is the minimum value of the cost function $d(\pi) = \min cost$. Note that $Cost$ must be greater or equal to any lower bound, this last constraint avoids unnecessary job.

$$\begin{aligned}&distance\_cop(\pi, N, Model) \text{ :-}\\&\quad bound(\pi, Model, LowerBound, UpperBound),\\&\quad length(B, UpperBound),\\&\quad upperbound\_constraint(\pi, B, UpperBound),\\&\quad sum(B, Cost),\\&\quad Cost \geq LowerBound,\\&\quad minimize(Cost, N).\end{aligned} \tag{6}$$

The *upperbound_constraint* predicate (7) retrieves the value of B for every transposition ρ_k and inserts the ρ_k effects on permutations sequence. An important constraint is check if it is possible to sort a permutation using the remaining amount of transposition, this constraint avoids unnecessary calculus.

$$\begin{aligned}&upperbound\_constraint(\iota, [\,], \_UpperBound).\\&upperbound\_constraint(\pi, [B|Bs], UpperBound) \text{ :-}\\&\quad transposition\_cop(\pi, \sigma, \_I, \_J, \_K, B),\\&\quad bound(\pi, Model, LowerBound, \_UpperBound),\\&\quad UpperBound \geq LowerBound,\\&\quad upperbound\_constraint(\sigma, Bs, UpperBound - 1).\end{aligned} \tag{7}$$

As we did to CSP models, all the COP models have the above structure. We used the COP models to analyse the upper bounds on Lemmas 5 and 6. We call **cg_cop** the model that uses the cycle graph upper bound (Lemma 5) and **gg_cop** the model that uses the Γ-graph upper bound (Lemma 6). For both models we used the cycle graph lower bound (Lemma 4). As we did for the CSP models, we create for comparison purposes an **def_cop** model where the bound is the permutation size.

4 Computational Experiments

All the constraint logic programming models were implemented using the open source programming system *Eclipse* [11]. We recommend Apt and Wallace [1] and Marriott and Stuckey [21] as introduction for CLP using *Eclipse*. We made experiments on a Intel® Xeon™, 3.20GHz computer, with 3.5 GB RAM, Ubuntu 7 and Eclipse 6.0.

Table 1 summarizes our results. The CPU time reported here refers to an average of all permutation π, if $|\pi| < 7$, otherwise it is an average of 1000 instances chosen by random. These times are given in seconds and grow very fast as the instance size increases. This behavior is due to the exponential search space.

Table 1. Average time (in seconds) for the computation of transposition distance between random permutations and the identity. A symbol "–" is printed whenever a model could not finish the entire tests within a time limit of 15 hours.

	CSP				COP		
Size	**def_csp**	**br_csp**	**cg_csp**	**cg_csp_red**	**def_cop**	**cg_cop**	**gg_cop**
4	0.006	0.010	0.005	0.005	0.134	0.019	0.039
5	0.069	0.087	0.009	0.006	2.530	0.061	0.149
6	1.887	2.367	0.022	0.013	–	1.842	4.421
7	51.69	30.707	0.045	0.024	–	3.797	39.024
8	–	–	0.233	0.104	–	–	–
9	–	–	0.946	0.313	–	–	–
10	–	–	6.816	2.016	–	–	–
11	–	–	20.603	4.212	–	–	–

Table 1 shows that cycle graph are a better approach than breakpoints. It is an expected result because the cycle graph lower bound is tighter than the breakpoint lower bound. Another expected result is that the reduction by breakpoints is a very useful approach, for $|\pi| = 11$ the average CPU time of **cg_csp_red** is almost 4.9 times better than **cg_csp** that don't use it.

The drawback of **cg_csp_red** and any other model that uses reduction is that it is harder to obtain the sequence of transposition that ordered the input permutation. The other approaches need almost no modifications to retrieve the transposition sequence, it is only necessary to add an extra list of variables.

The COP models have the worst CPU time. However, it is important to mention that we did not create any search heuristic that could improve the results. We used a generic solver to constrained optimisation problem that realises the basic branch and bound algorithm augmented with the constraint propagation. For this reason, efforts are still necessary to conclude that CSP models are the best approach based only on experimental results.

However, the search mechanism of the COP models will always use the strategy of finding a solution and try to improve it by looking for new solutions with improved value of the cost function. In general, this mechanism leads to a search space greater than the CSP models that uses a bottom-up search.

For example, let A be a lower bound and B be an upper bound for a permutation π, $d(\pi) = t$, and $A \leq t \leq B$. The COP models will produce better and better solutions $\rho_1 \dots \rho_l$ such as $l \in [t + 1..B]$ until it finds an optimal solution with size t. But the search mechanism needs to prove that this solution is optimal, so it will search on sequences of transpositions of size $l \in [A..t-1]$. In other words, the search space of COP has the sequences of transpositions with size in $[A..B]$ and the bottom-up strategy of CSP models lead to a search space with the sequences of transpositions with size in $[A..t]$.

CSP models even with the simple approaches that we presented here are very efficient. However, some improvements could be used. For example, given a permutation π, the cycle graph $G(\pi)$ gives information about what transpositions increase the number of cycles and the number of odd cycles. This information could "cut" the search space making the search faster. It also could improve the variable ordering on the search tree, because transpositions that increases the number of odd cycles have greater chance of being in an optimal sequence.

Finally, our best model **cg_csp_red** is much better than the ILP models of Dias and Souza [9] where the events are restricted to transpositions. They reported tests using 100 instances randomly generated and with $|\pi| = 9$ they found solutions with 143.5 seconds on average. Our model finds solution with 0.313 second on average. With $|\pi| = 10$ their models become prohibitive.

Further analysis between our models and the models created by Dias and Souza [9] are not possible because the instances and the computational environment used in each test are different.

5 Conclusion and Future Works

In this paper we presented Constraint Logic Programming models for the sorting by transposition problem and compared our experimental results with the Dias and Souza [9] polynomial-sized Integer Linear Programming models where the events are restricted to transpositions. Our analysis shows that we achieved better performance.

For future works we plan to use the toric equivalences [13] to decrease the amount of permutations we must to search in order to obtain the distance. Using toric equivalences some permutations will not be branched, because the cycle graph associated with them are isomorphic [18] with the cycle graph of another permutation already branched.

For example, the seven following permutations are in the same toric group: (1 6 2 4 5 3), (1 6 3 4 2 5), (2 3 1 5 6 4), (3 5 6 4 1 2), (4 5 3 6 1 2), (5 1 3 4 2 6), (5 2 3 1 4 6). When one of the permutations are branched, the transposition distance for the entire group is calculated. However, our models recalculate the distance as soon as it is required for the other permutations. We think that toric equivalences combined with breakpoint reduction are powerful tools to generate better models.

Another improvement would be to use the cycle graph of a permutation in order to choose a set of transpositions that will be branched before the others. We know some transpositions have more chance of being in an optimal sequence [3]. However, we did not pay attention to this issue and, given a permutation, we applied the transpositions $\rho(1,2,3)$, $\rho(1,2,4)$, $\rho(1,2,5)$ and so on.

Bafna and Pevzner [3] proposed a classification of transpositions in many subsets like valid 2-moves, 2-moves, good 0-moves, valid 0-moves and 0-moves based on cycle graph configurations before and after the transposition. Most of time, if exist valid 2-moves for a permutation π, than they are preferred. Thus, we plan to give priority to moves with more chance of being in an optimal sequence.

Another direction of future research is to model distance problems where events are restricted to reversals [2], or to a combination of reversals and transpositions.

References

1. Apt, K., Wallace, M.: Constraints Logic Programming using Eclipse. Cambridge (2007)
2. Bafna, V., Pevzner, P.A.: Sorting by reversals: Genome rearrangements in plant organelles and evolutionary history of X chromosome. Molecular Biology and Evolution 12(2), 239–246 (1995)
3. Bafna, V., Pevzner, P.A.: Sorting by Transpositions. SIAM Journal on Discrete Mathematics 11(2), 224–240 (1998)
4. Benoît-Gagné, M., Hamel, S.: A New and Faster Method of Sorting by Transpositions. In: Ma, B., Zhang, K. (eds.) CPM 2007. LNCS, vol. 4580, pp. 131–141. Springer, Heidelberg (2007)
5. Caprara, A., Lancia, G., Ng, S.-K.: A Column-Generation Based Branch-and-Bound Algorithm for Sorting by Reversals. DIMACS Series in Discrete Mathematics and Theoretical Computer Science, The American Mathematical Society 47, 213–226 (1999)
6. Caprara, A., Lancia, G., Ng, S.-K.: Sorting Permutations by Reversals through Branch-and-Price. Technical Report OR-99-1, DEIS - Operations Research Group, University of Bologna (1999)
7. Caprara, A., Lancia, G., Ng, S.-K.: Fast Practical Solution of Sorting by Reversals. In: Proceedings of the 11th ACM-SIAM Annual Symposium on Discrete Algorithms (SODA 2000), San Francisco, USA, pp. 12–21. ACM Press, New York (2000)
8. Christie, D.A.: Genome Rearrangement Problems. PhD thesis, Glasgow University (1998)
9. Dias, Z., Souza, C.: Polynomial-sized ILP Models for Rearrangement Distance Problems. In: BSB 2007 Poster Proceedings (2007)
10. Dobzhansky, T., Sturtevant, A.H.: Inversions in the third chromosome of wild races of *Drosophila pseudoobscura*, and their use in the study of the history of the species. Proceedings of the National Academy of Science 22, 448–450 (1936)
11. The Eclipse Constraint Programming System (March 2009), `http://www.eclipse-clp.org`
12. Elias, I., Hartmn, T.: A 1.375-Approximation Algorithm for Sorting by Transpositions. IEEE/ACM Trans. Comput. Biol. Bioinformatics 3(4), 369–379 (2006)
13. Eriksson, H., Eriksson, K., Karlander, J., Svensson, L., Wästlund, J.: Sorting a Bridge Hand. Discrete Math. 241(1-3), 289–300 (2001)
14. Fortuna, V.J.: Distâncias de transposição entre genomas. Master's thesis, Institute of Computing, University of Campinas (2005)

15. Hannenhalli, S., Pevzner, P.A.: Transforming Cabbage into Turnip (Polynomial Algorithm for Sorting Signed Permutations by Reversals). In: Proceedings of the Twenty-Seventh Annual ACM Symposium on the Theory of Computing, Las Vegas, USA, May 1995, pp. 178–189 (1995)
16. Hannenhalli, S., Pevzner, P.A.: Transforming Men into Mice (Polynomial Algorithm for Genomic Distance Problem). In: Proceedings of the 36th Annual Symposium on Foundations of Computer Science (FOCS 1995), October 1995, pp. 581–592. IEEE Computer Society Press, Los Alamitos (1995)
17. Hartman, T., Sharan, R.: A Simpler 1.5-approximation Algorithm for Sorting by Transpositions, pp. 156–169. Springer, Heidelberg (2003)
18. Hausen, R.A., Faria, L., Figueiredo, C.M.H., Kowada, L.A.B.: On the toric graph as a tool to handle the problem of sorting by transpositions. In: Bazzan, A.L.C., Craven, M., Martins, N.F. (eds.) BSB 2008. LNCS (LNBI), vol. 5167, pp. 79–91. Springer, Heidelberg (2008)
19. Kececioglu, J.D., Ravi, R.: Of Mice and Men: Algorithms for Evolutionary Distances Between Genomes with Translocation. In: Proceedings of the 6th Annual Symposium on Discrete Algorithms, January 1995, pp. 604–613. ACM Press, New York (1995)
20. Labarre, A.: New Bounds and Tractable Instances for the Transposition Distance. IEEE/ACM Trans. Comput. Biol. Bioinformatics 3(4), 380–394 (2006)
21. Marriott, K., Stuckey, P.J.: Programming with Constraints: An Introduction. MIT Press, Cambridge (1998)
22. Mira, C.V.G., Dias, Z., Santos, H.P., Pinto, G.A., Walter, M.E.: Transposition Distance Based on the Algebraic Formalism. In: Bazzan, A.L.C., Craven, M., Martins, N.F. (eds.) BSB 2008. LNCS (LNBI), vol. 5167, pp. 115–126. Springer, Heidelberg (2008)
23. Palmer, J.D., Herbon, L.A.: Plant mitochondrial DNA evolves rapidly in structure, but slowly in sequence. Journal of Molecular Evolution 27, 87–97 (1988)
24. Walter, M.E.M.T., Dias, Z., Meidanis, J.: A New Approach for Approximating the Transposition Distance. In: Proceedings of the String Processing and Information Retrieval (SPIRE 2000) (September 2000)

Comparison of Spectra in Unsequenced Species

Freddy Cliquet[1,2], Guillaume Fertin[1], Irena Rusu[1], and Dominique Tessier[2]

[1] LINA, UMR CNRS 6241 Université de Nantes,
2 rue de la Houssinière, 44322, Nantes, Cedex 03, France
{freddy.cliquet,guillaume.fertin,irena.rusu}@univ-nantes.fr
[2] UR1268 BIA, INRA, Rue de la Géraudière, BP 71627, 44316 Nantes, France
dominique.tessier@nantes.inra.fr

Abstract. We introduce a new algorithm for the mass spectrometric identification of proteins. Experimental spectra obtained by tandem MS/MS are directly compared to theoretical spectra generated from proteins of evolutionarily closely related organisms. This work is motivated by the need of a method that allows the identification of proteins of unsequenced species against a database containing proteins of related organisms. The idea is that matching spectra of unknown peptides to very similar MS/MS spectra generated from this database of annotated proteins can lead to annotate unknown proteins. This process is similar to ortholog annotation in protein sequence databases. The difficulty with such an approach is that two similar peptides, even with just one modification (i.e. insertion, deletion or substitution of one or several amino acid(s)) between them, usually generate very dissimilar spectra. In this paper, we present a new dynamic programming based algorithm: PacketSpectralAlignment. Our algorithm is tolerant to modifications and fully exploits two important properties that are usually not considered: the notion of inner symmetry, a relation linking pairs of spectrum peaks, and the notion of packet inside each spectrum to keep related peaks together. Our algorithm, PacketSpectralAlignment is then compared to SpectralAlignment [1] on a dataset of simulated spectra. Our tests show that PacketSpectralAlignment behaves better, in terms of results and execution time.

1 Introduction

In proteomics, tandem mass spectrometry (MS/MS) is a general method used to identify proteins. At first, during the MS/MS process, the peptides issued from the digestion of an unknown protein by an enzyme are ionized so that their mass may be measured. Then, the mass spectrometer isolates each peptide and fragments it into smaller ions, before measuring the corresponding masses. This process provides for each peptide an **experimental spectrum** in the form of a series of peaks, each peak corresponding to a mass that has been measured. From these experimental spectra, we aim at retrieving the corresponding peptide sequences. Then, by combining the peptide sequences from different spectra, our goal is to relate the unknown protein to one of the proteins stored in a protein

K.S. Guimarães, A. Panchenko, T.M. Przytycka (Eds.): BSB 2009, LNBI 5676, pp. 24–35, 2009.

database such as SwissProt. Given an experimental spectrum, there are two main possibilities to obtain the corresponding peptide sequence: *de novo* sequencing or *spectra comparison.*

***De novo*:** In *de novo* sequencing, the spectrum is directly interpreted and a corresponding amino acid sequence of the peptide is inferred without using any other data than the spectrum itself. Currently, the analysis of unsequenced species is mainly done by *de novo* interpretation [2,3]. The main reason is the speed of each interpretation and the fact that the peptide sequence does not need to be part of a protein database. An important drawback is that this method requires high quality spectra and still leads to a lot of interpretation errors [4,5].

Spectra Comparison: In spectra comparison, an experimental spectrum is compared with theoretically predicted spectra. The theoretical spectra are inferred from the different peptides generated by an *in-silico* digestion of all the proteins contained in a database. A score function is used to evaluate each possible comparison, and the results are ordered according to this score. The theoretical spectrum with the best score is associated to the corresponding experimental spectrum. There are a number of existing softwares that match uninterpreted MS/MS experimental spectra to theoretical spectra, such as SEQUEST [6] or MASCOT [7]. They are based on the Shared Peaks Count (SPC) algorithm, an algorithm that simply counts the number of peaks in common between the two spectra, but presents zero tolerance to **modifications** (i.e. insertion, deletion or substitution of one or several amino acid(s)). This is due to the fact that the slightest modification in a peptide sequence highly changes the contents of the corresponding spectrum, and thus can lead to false identifications. Two approaches have already been explored to take modified peptides into account. The first one consists in extending the database by applying all the possible modifications to each peptide of the base. However, this solution, leading to an exponential number of possibilities, is of course too time consuming [8]. The other one, SpectralAlignment [1,9,10], is a dynamic programming algorithm that has been designed to identify peptides even in presence of modifications. This method works rather well for one or two modifications, but for a larger number of modifications, SpectralAlignment is not really sustainable [9].

Spectra comparison has an essential advantage, namely the precision in the comparison, allowing information to be drawn even from spectra which used to be unexploitable with a *de novo* approach. But spectra comparison has a major downfall: the running time, that is highly dependent on the protein database size used to infer theoretical spectra.

Given that the proteins we are trying to identify come from unsequenced species, the idea is to find similar proteins on phylogenetically related organisms. This is why our method will have to allow the correspondence of a spectrum with a slightly different peptide. Yet, we need results with few errors and for most spectra. Although *de novo* is specially designed to treat unsequenced species, it still leads to lots of misinterpreted or uninterpreted spectra. That is why the development of a new spectra comparison method tolerant to modifications appears interesting for us.

2 Definitions and Notations

An MS/MS spectrum is obtained by selection, isolation and fragmentation of peptides within the mass spectrometer. Each peak results from the dissociation of a given peptide into two fragmented ions: an N-terminal one, and a C-terminal one. The **N-terminal** ions (that are called a, b, c, see for instance Figure 1) represent the left part of the peptide sequence, and the **C-terminal** ions (that are called x, y, z) represent the right part of the peptide sequence. The position where the fragmentation appears is located around the junction of two amino acids, as described in Figure 1. The b and y ions mark the exact junction point. In the rest of this paper, we will say that two different ions are **dependent** if they are issued from the fragmentation between the same two successive amino acids inside a peptide. For instance, in Figure 1, for a given $i \in [1;3]$, a_i, b_i, c_i, x_i, y_i and z_i are mutually dependent ions. In a spectrum, a peak corresponding to a C-terminal (resp. N-terminal) ion is called C-terminal (resp. N-terminal). Moreover, peaks corresponding to dependent ions are called dependent peaks.

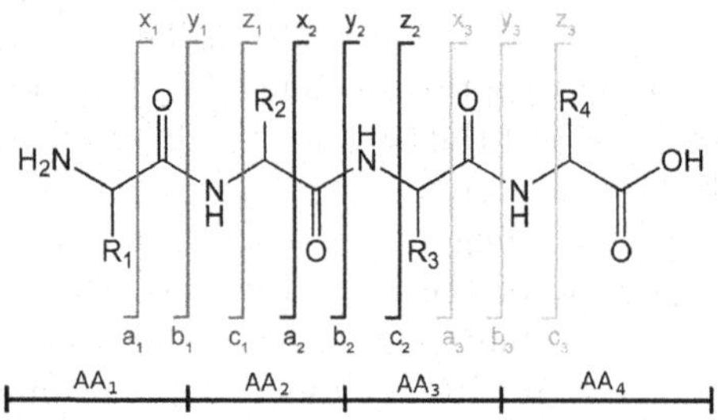

Fig. 1. This figure shows the fragmentation points inside a peptide containing four amino acids (AA_i with $i \in [1;4]$). R_i ($i \in [1;4]$) are chemical compounds that determine the corresponding amino acid. In this example, there are three sets of dependent ions.

We can notice that a symmetry exists between N-terminal and C-terminal peaks for a given fragmentation. In Figure 2 (a), the N-terminal peak located at position $m(GL)$ and the C-terminal peak located at position $m(MPRG)$ are linked by the relation $m(MPRG) = m(GLMPRG) - m(GL) - 20$ (-20 is due to the fact that the peptide is not symmetric at its extremities, see Figure 1, and to the ionization). This is a critical notion that is valid for any fragmentation and is seldom used. We call this relation **inner symmetry**.

Any spectrum produced by a mass spectrometer is called an **experimental spectrum** (S_e), and any spectrum predicted *in-silico* from a peptide is called a **theoretical spectrum** (S_t). In the following figures, spectra will always display all of the nine most frequent peaks coming from the fragmentation [11,12].

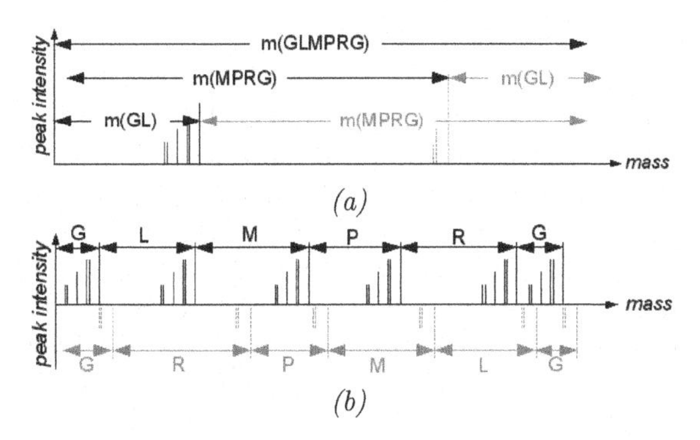

Fig. 2. (a) This piece of a spectrum shows the peaks created when the peptide GLMPRG (of mass $m(GLMPRG)$) is broken into peptides GL and MPRG. **(b)** is the spectrum of the peptide GLMPRG (for a better visualization, the N-terminal peaks are above the axis, the C-terminal ones are below the axis).

Considering that the masses where the peaks appear can be seen as integers, we can represent a spectrum by a vector of booleans. This vector contains, for every mass, a boolean attesting the presence of a peak ('true') or its absence ('false'). Vector V_t represents the spectrum S_t and V_e represents S_e. Then, as in the case of sequences, we can align the elements of V_t and V_e two by two while allowing the insertion of gaps. It is only necessary to ensure that both vectors have the same length and that a gap is not aligned with another gap. In this representation, a **shift** corresponds to the insertion of a gap in either V_t or V_e, which itself corresponds to a peptide modification.

A **score** is used to evaluate an alignment, which represents the similarity between both spectra, in the sense that a higher score means a higher similarity. For instance, the number of common peaks in a given alignment is a possible score. When transposed to V_t and V_e, this corresponds to the number of pairs of booleans in which both values are 'true' at the same position. In this context, the alignment having the highest score will be considered as the best alignment.

Our goal is, given an experimental spectrum, to compare it to all the theoretical spectra from the database. For each comparison, we want to find the best alignment.

3 Our Method

Because we want to align a spectrum S_e originated from unsequenced species with a spectrum S_t generated from phylogenetically related organisms, our method must be able to take modifications into account. Here, modification means insertion, deletion or substitution of one or several amino acids, but we will see that our algorithm, by nature, is able to handle other types of modifications such as post translational modifications.

Dynamic programming is an appropriate algorithmic method to find the best alignment between two spectra, even in presence of modifications. It has been used in this context by methods such as SpectralAlignement (SA) [1].

For our method to be tolerant to substitutions, we must be careful about the way we shift peaks when we want to improve our alignment, because a substitution, by changing the mass of an amino acid, not only changes one peak, but also some other peaks inside the whole spectrum. To take into account a substitution, a naïve algorithm could simply shift all peaks positioned after the location of the substituted amino acid. However, this could drastically change the positions of the peaks in the whole spectrum, as shown in Figure 3. It can be seen that such a modification destroys the link between N-terminal and C-terminal peaks, causing the inner symmetry to be broken. Moreover, the loss of information due to this rupture in the symmetry grows with the number of modifications.

In a previous work [1], Pevzner et al. considered this notion of symmetry by proposing, for each peak of each spectrum, to add its *symmetric twin* peak. However, they noticed two major drawbacks to this:

- The addition of noise (the symmetric twin of a noise peak is another noise peak).
- During the alignment, if a peak P (resp. a symmetric twin peak TP) is aligned, then TP (resp. P) must not be used in the alignment. Deciding, for each peak of the spectrum, which one of these peaks (P or TP) must be aligned, is an NP-complete problem [11].

But it is important to note that the two spectra S_e and S_t do not present the same properties. By construction, in S_t, we can identify each peak as an N-terminal or a C-terminal peak. This extra information is fully exploited in our algorithm and eliminates both previously raised drawbacks.

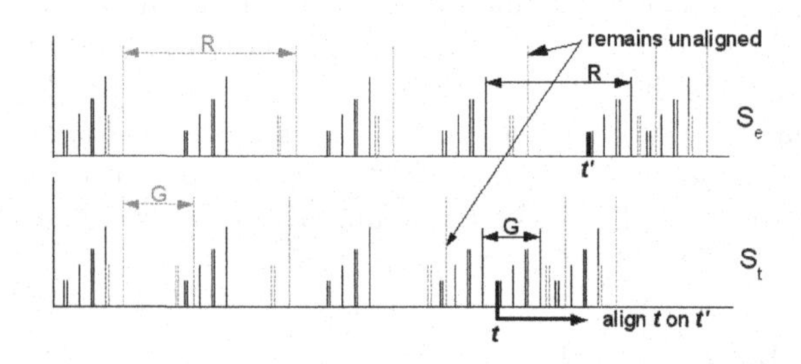

Fig. 3. In this figure, two spectra are represented. C-terminal peaks are dashed gray, N-terminal peaks are black. Spectrum S_e represents the peptide GLMPRG, spectrum S_t represents the peptide GLMPGG. This figure shows how one substitution in a peptide can highly change the contents of a spectrum (as it is, there are only 27 peaks aligned between S_e and S_t out of 54), and how shifting peaks from position t to position t' is not sufficient to correctly align S_e with S_t (the black arrow shows the shift, resulting in an alignment of 45 peaks between S_e and S_t out of 54). Although the shift improves the alignment, some of the peaks remain unaligned after the shift has been applied.

3.1 Symmetry

Theoretical Spectrum: We build the spectrum *in-silico*, so the location of all the different peaks is known. Considering this, we can easily remove all the C-terminal peaks and replace them by their respective symmetric peaks. Let m_i be the mass of the i-th C-terminal peak; its mass will be replaced by $M_{peptide} - m_i$ where $M_{peptide}$ is the mass of the peptide represented in S_t. The theoretical spectrum, after the application of the symmetry, is called **theoretical symmetric spectrum** (SS_t). Figure 4 (above) shows the spectrum S_t corresponding to the peptide *GLMPRG*. Figure 4 (below) shows the spectrum SS_t, that is the spectrum S_t to which the symmetry has been applied. Note that in SS_t, the distance between the N-terminal and C-terminal peaks from the same fragmentation is a constant (i.e. in spectrum SS_t from Figure 4, peaks t and t' (resp. u and u') are dependent and the distance between t and t' is the same than between u and u'), thus a shift of all the peaks positioned after a modification will still respect the *inner symmetry*. We also point out that our construction of SS_t is different that the one proposed by Pevzner et al. in [1] on two points: (1) we do not apply symmetry on the N-terminal peaks and (2) when applying symmetry on C-terminal peaks, we remove the original peaks.

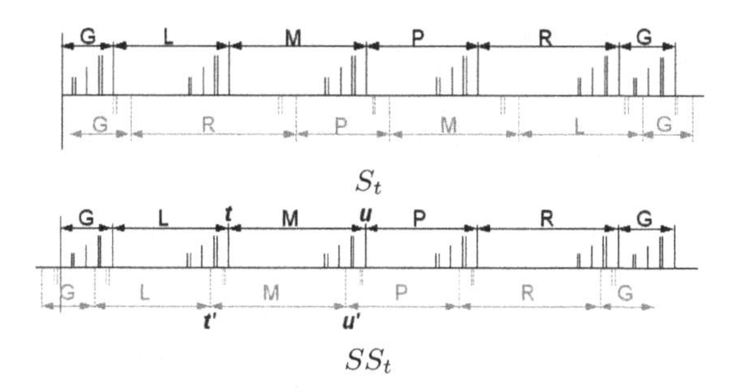

Fig. 4. S_t *(above)* represents the spectrum before any symmetry is applied. SS_t *(below)* represents the spectrum after the symmetry has been applied. In S_t the dashed peaks are the N-terminal peaks. In SS_t the dashed peaks are the N-terminal peaks to which symmetry has been applied.

Experimental Spectrum: As we do not know the ion type represented by each peak (in fact, a peak can represent the superposition of different ions of the same mass), we create a new symmetric peak for each existing peak of the spectrum. These peaks are created at position $M_{peptide} - m_i$, where $M_{peptide}$ represents the mass of the peptide measured by the mass spectrometer and m_i the mass of the i-th peak of S_e. The experimental spectrum, after the application of the symmetry is called **experimental symmetric spectrum** (SS_e). Figure 5 (above) shows the spectrum S_e corresponding to the peptide *GLMPGG*. Figure 5 (below) shows the spectrum SS_e, that is the spectrum S_e to which the symmetry has been applied.

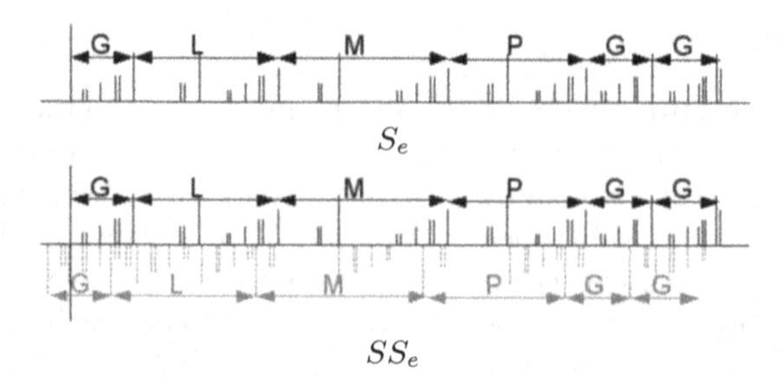

Fig. 5. S_e *(above)* represents the spectrum before any symmetry is applied. Then SS_e *(below)* represents the spectrum after the symmetry has been applied. In SS_e, the dashed peaks are S_e peaks to which symmetry has been applied.

During the alignment, we need to forbid the alignment of some pairs of peaks. For instance, N-terminal peaks from SS_t should be aligned with the original peaks from SS_e. This is important to guarantee that the solution is feasible. To do this, it is sufficient to keep track, for each peak, if it is a original peak or not.

3.2 Packet

The random fragmentation along the peptide backbone creates a number of different types of peaks. The location where the fragmentation occurs implies various peaks, a, b, c and x, y, or z (see Figure 1). Additionally, peaks corresponding to neutral loss (water, ammonia) are also frequently observed. For the construction of SS_t we choose to keep, for each type of fragmentation, the nine most frequent peaks observed in experimental spectra when using a *Quadrupole Time-of-Flight* mass spectrometer [11,12] (see also Figure 6). After the application of symmetry, the notion of inner symmetry does not depend anymore of the peptide mass, thus these nine peaks may be clustered in a single **packet**. A packet represents all the fragmentations occurring between two consecutive amino acids of the peptide, thus SS_t can now be represented by a group of packets. An example of packet is shown in Figure 6. Point R_p marks the **reference point** of a packet p, and will be referred to when we will talk about a *packet position.* Introducing this notion of indivisible packet into the comparison between experimental and theoretical spectra allow us to forbid any translation that pulls apart dependent peaks: indeed, a shift can only exist between two packets.

In addition, to align SS_t with SS_e, for each packet p of SS_t, R_p is positionned on a mass m of SS_e. This gives an alignment of score s between p and SS_e. If s goes past a threshold, then the mass m is considered as one of the **possible masses**. In the rest of this paper, this score will be the number of aligned peaks. Increasing the threshold T will speed up the alignment process because the number of possible masses will decrease (see Table 1).

Another constraint which forbids the overlapping of two packets is added; it represents the fact that an amino acid has a minimum mass. That way, we do not allow the algorithm to make small and unrealistic shifts just to slightly improve its score (something which happens very often with SA).

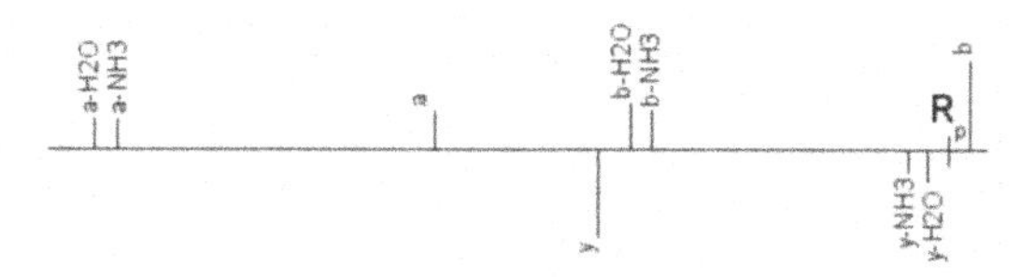

Fig. 6. The nine peaks representing the nine most current peaks that occur in MS/MS spectra (a, b, y and some of their variant peaks that are due to water or ammonia loss) result after the symmetry has been applied, to this packet. This packet is particularly suited for *Quadrupole Time-of-Flight* mass spectrometer.

Note that it is possible to modify the contents of a packet in order to adapt this notion for other types of mass spectrometers.

3.3 PacketSpectralAlignment Algorithm

Our PacketSpectralAlignment (PSA) method needs three parameters: (i) SS_e as described in Section 3.1, (ii) SS_t for which the peaks are clustered into packets as described in Section 3.2 and (iii) K, the maximum number of allowed shifts. Our PSA algorithm (Algorithm 1) uses two matrices M and D. The value $M[p][m][k]$ represents the best score obtained so far for the alignment of the peaks of the first packets of SS_t, up to packet number p, with all the first peaks from SS_e, up to the alignment of R_p, with mass m, and with at most k shifts. The value $D[p][m][k]$ represents the best score of an alignment that must have R_p aligned with mass m in SS_e and containing at most k shifts. PSA will compute, for each possible number k of shifts, each possible mass m in SS_e and for each packet p of SS_t, the values $M[p][m][k]$ and $D[p][m][k]$.

The variable *best* contains the score of the best alignment met on the D matrix that could be extended with the current position (p, m, k) without more shifts (see Figure 7 for an illustration of how one of these alignments is found).

The *Score* function will return the score resulting from the alignment of the peaks of the p-th packet of SS_t with the peaks of SS_e, when R_p is positioned at the mass m of SS_e.

The D matrix is updated by choosing the best possibility between the two following cases:

a. we used the last value met on the diagonal, meaning no shift is needed, or

b. we must apply a shift and take the last score met on the diagonal in the $(k-1)$-th dimension of the matrix.

Then the M matrix is updated by taking the best alignment found until this point.

As an illustration, applying our algorithm PSA on the theoretical spectrum SS_t of Figure 4 and the experimental spectrum SS_e of Figure 5 gives a "perfect" alignment of the peaks (i.e. that is, 54 peaks out of 54 that are aligned) with one shift (corresponding to the substitution of the fifth amino acid R by G in SS_t).

Algorithm 1. PSA(ExperimentalSpectrum SS_e, TheoreticalSpectrum SS_t, Integer K)

Ensure: The best alignment between SS_e and SS_t with a max of K shifts
1: **for** $k = 0$ to K **do**
2: **for all** Possible masses m from SS_e **do**
3: **for all** Packets p from SS_t **do**
4: $best = \max \{D[p'][m - (R_p - R_{p'})][k] \mid p' < p\}$ /**(see Figure 7)**/
5: $s =$ Score(p,m)
6: $D[p][m][k] = \max(best{+}s, M[p{-}1][m{-}PacketSize][k{-}1]{+}s)$ /*$PacketSize$ *is the constant representing the size of a packet (i.e. the distance between the first and the last peak of a packet)**/
7: $M[p][m][k] = \max(D[p][m][k], M[p-1][m][k], M[p][m-1][k])$
8: **end for**
9: **end for**
10: **end for**
11: **return** $M[NbPacket][MAX][K]$ /*$NbPacket$ *is the number of packets composing* SS_t *and* MAX *is the highest mass represented by a peak inside* SS_e*/

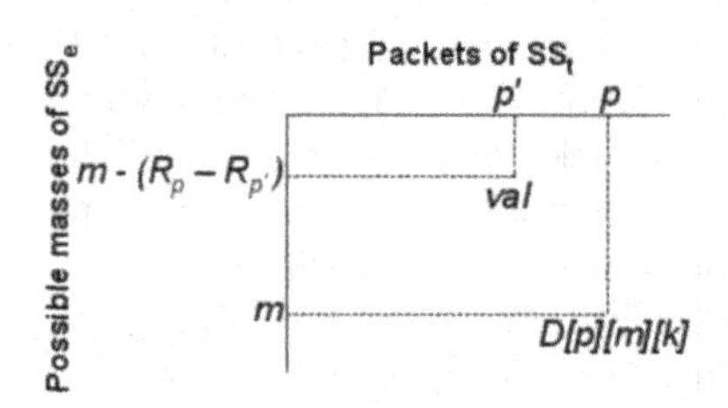

Fig. 7. This figure shows how to find a value $val = D[p'][m - (R_p - R_{p'})][k]$, with $p' < p$

4 Results

We compare our algorithm to SA on a set of simulated data. We generate a dataset of 1000 random peptides of random size in [10, 25] in order to constitute a database that will be used to create the theoretical symmetric spectra. Each peptide in the database is then modified in 5 different versions by applying 0 to 4 random substitutions of amino acids. These modified peptides are used to create 5 sets of artificial experimental symmetric spectra (one for each different number of modifications). These spectra are constituted using the nine most frequents peaks that are created considering the probability of apparition observed by Frank et al. [12]. Noise has been introduced in each spectra, adding 50% more peaks at random masses. All tests have been made using 1 *dalton* precision. For each SS_e, we call the **target peptide** of SS_e (denoted $TP(SS_e)$) the original peptide sequence from the dataset that has been modified to obtain the spectrum.

Each SS_e is compared with each SS_t. The score (here, the number of common peaks in the best alignment) resulting from each comparison is memorized

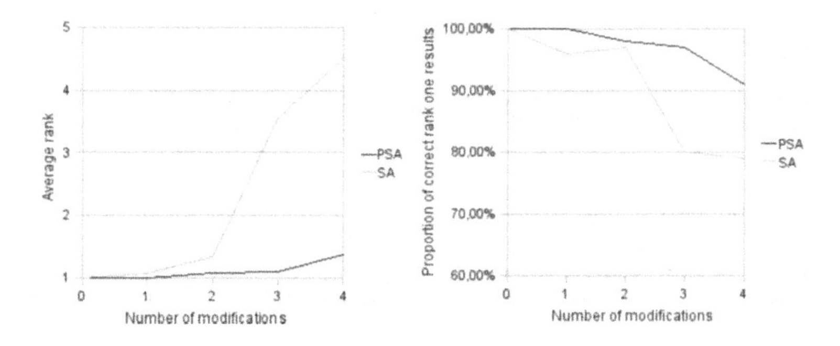

Fig. 8. Comparison of SA and PSA on our sets of 1000 random peptides

and used to order the set of peptides for each SS_e. In the ideal case, for an experimental spectrum SS_e, $TP(SS_e)$ should obtain the highest score and thus should have rank one. Thus, by looking at the rank of the target, we can evaluate the capacity for PSA to take modifications into account. That is why we use the average rank of the target peptide, as well as the proportion of target peptides having rank one, as indicators to evaluate the two algorithms (as shown in Figure 8).

During the comparisons, the parameters used by SA and our algorithm PSA are the same. In particular, we use the same score function, which is the number of common peaks in both spectra. The number of shifts used by these two methods is set dynamically, depending on the size of the two compared spectra. We could have fixed this to a constant value, but allowing for instance $\frac{N}{2}$ shifts in an N amino acids long peptide does not make any sense, so we chose to allow k shifts for a peptide of mass M where $k = \lceil \frac{M}{600} \rceil + 1$. In the case of PSA, the threshold T used to determine the possible masses kept in symmetric experimental spectra is set to 2.

Our tests show that the two algorithms have a comparable behaviour for 0 to 2 shifts, with a slight advantage for our algorithm. However, for more than two shifts, SpectralAlignment presents a fast deterioration of its results, while PacketSpectralAlignment still gives good results (see Figure 8). We also note that on these tests, for a threshold T of 2, our algorithm PSA is twice as fast as SA.

We have also evaluated the benefits supplied by the packets, and more particularly by the number of possible masses. As explained in Section 3.2, we do not test all masses in SS_e, but only those masses m inducing an alignment of at least T peaks when the reference point R_p of a packet p from SS_t is positionned at mass m. To evaluate this, we have computed the number of possible masses for different values of T on four different datasets. The first one is a set of 1000 simulated spectra of size $[10, 25]$ with 50% of noise peaks, generated the same way as described at the beginning of Section 4. On this dataset, a spectrum contains on average 150 peaks. Then we use three sets of 140 experimental maize spectra on which we apply different filters: (1) we keep all peaks (meaning an average of 275 peaks per spectrum), (2) we keep the 100 most intense peaks, and (3) we keep the 50 most intense peaks. Table 1 shows the evolution of the

Table 1. Evaluation of the number of possible masses on four sets of spectra depending on the threshold T

Number of Possible Masses					
		Threshold T			
		1	2	3	4
Simulated spectra		485	134	39	14
Experimental Maize spectra	*no filtering*	689	312	141	61
	100 most intense peaks	540	180	57	17
	50 most intense peaks	346	79	18	4

number of possible masses in function of the threshold T for each set of spectra. We can notice that the number of possible masses decreases considerably when T is increased.

5 Conclusion

We have developed PacketSpectralAlignment, a new dynamic programming based algorithm that fully exploits, for the first time, two properties that are inherent to MS/MS spectra. The first one consists in using the *inner symmetry* of spectra and the second one is the grouping of all dependent peaks into *packets*. Although our algorithm was at first motivated by the identification of proteins in unsequenced organisms, it does not set any constraints on the allowed shifts in the alignment. Thus, PSA is also able to handle the discovery of post translational modifications.

Our results are very positive, showing a serious increase in peptides identification in spite of modifications. The sensibility has been significantly increased, while the execution time has been divided by more than two. More tests on experimental data will allow us to evaluate more precisely the benefits provided by our new algorithm. In the future, a better consideration of other points, such as spectra quality, will be added. Moreover, the score will be improved by taking into account other elements such as peaks intensity.

Acknowledgments. MS/MS experimental spectra were performed with the facilities of the platform Biopolymers, Interactions and Structural Biology, INRA Nantes. The authors thank Dr Hélène Rogniaux for fruitful discussions about MS/MS spectra interpretation. This research was supported by grant from the Region Pays de la Loire, France.

References

1. Pevzner, P.A., Dancík, V., Tang, C.L.: Mutation-tolerant protein identification by mass spectrometry. J. Comput. Biol. 7(6), 777–787 (2000)
2. Habermann, B., Oegema, J., Sunyaev, S., Shevchenko, A.: The power and the limitations of cross-species protein identification by mass spectrometry-driven sequence similarity searches. Mol. Cell. Proteomics 3(3), 238–249 (2004)

3. Grossmann, J., Fischer, B., Baerenfaller, K., Owiti, J., Buhmann, J.M., Gruissem, W., Baginsky, S.: A workflow to increase the detection rate of proteins from unsequenced organisms in high-throughput proteomics experiments. Proteomics 7(23), 4245–4254 (2007)
4. Pevtsov, S., Fedulova, I., Mirzaei, H., Buck, C., Zhang, X.: Performance evaluation of existing de novo sequencing algorithms. J. Proteome. Res. 5(11), 3018–3028 (2006)
5. Pitzer, E., Masselot, A., Colinge, J.: Assessing peptide de novo sequencing algorithms performance on large and diverse data sets. Proteomics 7(17), 3051–3054 (2007)
6. Eng, J., McCormack, A., Yates, J.: An approach to correlate tandem mass spectral data of peptides with amino acid sequences in a protein database. J. Am. Soc. Mass Spectrom 5(11), 976–989 (1994)
7. Perkins, D.N., Pappin, D.J., Creasy, D.M., Cottrell, J.S.: Probability-based protein identification by searching sequence databases using mass spectrometry data. Electrophoresis 20(18), 3551–3567 (1999)
8. Yates, J.R., Eng, J.K., McCormack, A.L., Schieltz, D.: Method to correlate tandem mass spectra of modified peptides to amino acid sequences in the protein database. Anal. Chem. 67(8), 1426–1436 (1995)
9. Pevzner, P.A., Mulyukov, Z., Dancik, V., Tang, C.L.: Efficiency of database search for identification of mutated and modified proteins via mass spectrometry. Genome Res. 11(2), 290–299 (2001)
10. Tsur, D., Tanner, S., Zandi, E., Bafna, V., Pevzner, P.A.: Identification of post-translational modifications by blind search of mass spectra. Nat. Biotechnol. 23(12), 1562–1567 (2005)
11. Dancik, V., Addona, T., Clauser, K., Vath, J., Pevzner, P.: De novo peptide sequencing via tandem mass spectrometry. Journal of Computational Biology 6(3-4), 327–342 (1999)
12. Frank, A.M., Savitski, M.M., Nielsen, M.L., Zubarev, R.A., Pevzner, P.A.: De novo peptide sequencing and identification with precision mass spectrometry. J. Proteome Res. 6(1), 114–123 (2007)

BiHEA: A Hybrid Evolutionary Approach for Microarray Biclustering

Cristian Andrés Gallo[1], Jessica Andrea Carballido[1], and Ignacio Ponzoni[1,2]

[1] Laboratorio de Investigación y Desarrollo en Computación Científica (LIDeCC), Departamento de Ciencias e Ingeniería de la Computación, Universidad Nacional del Sur, Av. Alem 1253, 8000, Bahía Blanca, Argentina
[2] Planta Piloto de Ingeniería Química (PLAPIQUI) - UNS – CONICET Complejo CRIBABB, Co. La Carrindanga km.7, CC 717, Bahía Blanca, Argentina
{cag,jac,ip}@cs.uns.edu.ar

Abstract. In this paper a new hybrid approach that integrates an evolutionary algorithm with local search for microarray biclustering is presented. The novelty of this proposal is constituted by the incorporation of two mechanisms: the first one avoids loss of good solutions through generations and overcomes the high degree of overlap in the final population; and the other one preserves an adequate level of genotypic diversity. The performance of the memetic strategy was compared with the results of several salient biclustering algorithms over synthetic data with different overlap degrees and noise levels. In this regard, our proposal achieves results that outperform the ones obtained by the referential methods. Finally, a study on real data was performed in order to demonstrate the biological relevance of the results of our approach.

Keywords: gene expression data, biclustering, evolutionary algorithms.

1 Introduction

The task of grouping genes that present a related behavior constitutes a growing investigation area into the research field of gene expression data analysis. The classification is performed according to the genes' expression levels in the Gene Expression Data Matrix (GEDM). The success in this task helps in inferring the biological role of genes. The study of these complex interactions constitutes a challenging research field since it has a great impact in various critical areas. In this context, the microarray technology arose as a fundamental tool to provide information about the behavior of thousands of genes. The information provided by a microarray experiment corresponds to the relative abundance of the mRNA of genes under a given condition. The abundance of the mRNA is a metric that can be associated to the expression level of the gene. This information can be arranged into a matrix, namely GEDM, where rows and columns correspond to genes and experiments respectively.

In most cases, during the process of detecting gene clusters, all of the genes are not relevant for all the experimental conditions, but groups of them are often co-regulated and co-expressed only under some specific conditions. This observation has led the attention to the design of biclustering methods that simultaneously group genes and

K.S. Guimarães, A. Panchenko, T.M. Przytycka (Eds.): BSB 2009, LNBI 5676, pp. 36–47, 2009.

samples [1]. In this regard, a suitable bicluster consists in a group of rows and columns of the GEDM that satisfies some similarity score [2] in union with other criteria.

In this context, a new multi-objective evolutionary approach for microarray biclustering is presented, which mixes an aggregative evolutionary algorithm with features that enhance its natural capabilities. To the best of our knowledge, this methodology introduces two novel features that were never addressed, or partially dealt-with, by other evolutionary techniques designed for this problem instance. The first contribution consists in the design of a recovery process that extracts the best solutions through the generations. The other new characteristic is the incorporation of an elitism procedure that controls the diversity in the genotypic space. The paper is organized as follows: in the next section some concepts about microarray biclustering are defined; then, a brief review on relevant existing methods used to tackle this problem is presented; in Section 4 our proposal is introduced; then, in Section 5, the experiments and the results are put forward; finally some conclusions are discussed.

2 Microarray Biclustering

As abovementioned, expression data can be viewed as a matrix **E** that contains expression values, where rows correspond to genes and columns to the samples taken at different experiments. A matrix element e_{ij} contains the measured expression value for the corresponding gene i and sample j. In this context, a bicluster is defined as a pair (G, C) where $G \subseteq \{1,\dots, m\}$ is a subset of genes (rows) and $C \subseteq \{1,\dots, n\}$ is a subset of conditions [2]. In general, the main goal is to find the largest bicluster that does not exceed certain homogeneity constrain. It is also important to consider that the variance of each row in the bicluster should be relatively high, in order to capture genes exhibiting fluctuating coherent trends under some set of conditions. The size $g(G,C)$ is the number of cells in the bicluster. The homogeneity $h(G,C)$ is given by the mean squared residue score, while the variance $k(G,C)$ is the row variance [2]. Therefore, our optimization problem can be defined as follows:
maximize

$$g(G,C) = |G||C| \quad (1)$$

$$k(G,C) = \frac{\sum_{g\in G, c\in C}(e_{gc} - e_{gC})^2}{|G|\cdot|C|} \quad (2)$$

subject to

$$h(G,C) \le \delta \quad (3)$$

with $(G,C) \in X$, $X = 2^{\{1,\dots,m\}} \times 2^{\{1,\dots,n\}}$ being the set of all biclusters, where

$$h(G,C) = \frac{1}{|G|\cdot|C|} \sum_{g\in G, c\in C}(e_{gc} - e_{gC} - e_{Gc} + e_{GC})^2 \quad (4)$$

is the mean squared residue score,

$$e_{gC} = \frac{1}{|C|}\sum_{c\in C} e_{gc}, \quad e_{Gc} = \frac{1}{|G|}\sum_{g\in G} e_{gc} \quad (5,6)$$

are the mean column and row expression values of (G,C) and

$$e_{GC} = \frac{1}{|G|\cdot|C|} \sum_{g\in G, c\in C} e_{gc} \ . \tag{7}$$

is the mean expression value over all the cells that are contained in the bicluster (G,C). The user-defined threshold $\delta>0$ represents the maximum allowable dissimilarity within the cells of a bicluster. In other words, the residue quantifies the difference between the actual value of an element e_{gc} and its expected value as predicted for the corresponding row mean, column mean, and bicluster mean. A bicluster with a mean square residue lower than a given value δ is called a δ-bicluster. The problem of finding the largest square δ-bicluster is NP-hard [2] and, in particular, Evolutionary Algorithms (EAs) are well-suited for dealing these problems [3, 4, 5].

3 GEDM: Main Biclustering Methods

Cheng and Church's Approach (CC): Cheng and Church [2] were the first to apply the concept of biclustering on gene expression data. Given a data matrix **E** and a maximum acceptable mean squared residue score ($h(G,C)$), the goal is to find subsets of rows and subsets of columns with a score no larger than δ. In order to achieve this goal, Cheng and Church proposed several greedy row/column removal/ addition algorithms that are then combined in an overall approach. The multiple node deletion method removes all rows and columns with row/column residue superior to $\delta.\alpha$ in every iteration, where α is a parameter introduced for the local search procedure. The single node deletion method iteratively removes the row or column that grants the maximum decrease of $h(G,C)$. Finally, the node addition method adds rows and columns that do not increase the actual score of the bicluster. In order to find a given number of biclusters, the approach is iteratively executed on the remained rows and columns that are not present in the previous obtained biclusters.

Iterative Signature Algorithm (ISA): The most important conceptual novelty of this approach [6] is the focus on the desired property of the individual co-regulated bicluster that is going to be extracted from the expression data matrix. According to the definition of the authors, such a transcription bicluster consists of all genes that are similar when compared over the conditions, and all conditions that are similar when compared over the genes. This property is referred as self-consistency. In this regard, they proposed to identify modules by iteratively refining random input gene sets, using the signature algorithm previously introduced by the same authors. Thus, self-consistent transcription modules emerge as fixed-points of this algorithm.

BiMax: The main idea behind the Bimax algorithm [7] consists in the use of a divide and conquer strategy in order to partition **E** into three submatrices, one of which contains only 0-cells and therefore can be ignored in the following. The procedure is then recursively applied to the remaining two submatrices **U** and **V**; the recursion ends if the current matrix represents a bicluster, i.e. contains only 1s. If **U** and **V** do not share any rows and columns of **E**, the two matrices can be processed independently from each other. Yet, if **U** and **V** have a set of rows in common, special care is necessary to only generate those biclusters in **V** that share at least one common column with **CV**. A drawback of this approach is that only works on binary data matrices. Thus, the results strongly depend on the accuracy of the discretization step.

Order Preserving Submatrix Algorithm (OPSM): Ben-Dor et al. [8] defined a bicluster as an order-preserving submatrix (OPSM). According to their definition, a bicluster is a group of rows whose values induce a linear order across a subset of the columns. The work focuses on the relative order of the columns in the bicluster rather than on the uniformity of the actual values in the data matrix. More specifically, they want to identify large OPSMs. A submatrix is order-preserving if there is a permutation of its columns under which the sequence of values in every row is strictly increasing. In this way, Ben-Dor et al. aim at finding a complete model with highest statistically significant support. In the case of expression data, such a submatrix is determined by a subset of genes and a subset of conditions, such that, within the set of conditions, the expression levels of all genes have the same linear ordering. As such, Ben-Dor et al. addressed the identification and statistical assessment of co-expressed patterns for large sets of genes, and considered that, generally, data contains more than one such pattern.

Evolutionary Approaches: The first reported approach that tackled microarray biclustering by means of an EA was proposed by Bleuler *et al.* [5]. In this work, the use of a single-objective EA, an EA combined with a LS strategy [2] and the LS strategy alone [2] are analyzed. In the case of the EA, one novelty consists in a form of diversity maintenance on the phenotype space that can be applied during the selection procedure. For the case of the EA hybridized with a LS strategy, whether the new individual yielded by the LS procedure should replace the original individual or not is considered. As regards the LS as a stand alone strategy, they propose a new non-deterministic version, where the decision on the course of execution is made according to some probability. In the work of Mitra and Banka [3], the first approach that implements a Multi-Objective EA (MOEA) based on Pareto dominancy is presented. The authors base their work on the NSGA-II, and look for biclusters with maximum size and homogeneity. A LS strategy is applied to all of the individuals at the beginning of every generational loop. Finally, Gallo *et al.* [9] presents the $\mathrm{SPEA2}^{\mathrm{LS}}$, another MOEA combined with a LS [2] strategy. In this case, the authors base the algorithm on the SPEA2 [10], and seek biclusters with maximum rows, columns, homogeneity and row variance. A novel individual representation to consider the inverted rows of the data matrix is introduced. Also, a mechanism for re-orienting the search in terms of row variance and size is provided. The LS strategy is applied to all of the individuals in the resultant population of each generation.

4 BiHEA: Biclustering via a Hybrid Evolutionary Algorithm

The aim of our study is to use an evolutionary process to generate near optimal biclusters with coherent values following an additive model, according to the classification given by [1]. Thus, the EA is used to globally explore the search space X. However, it was observed that, in the absence of local search, stand-alone single-objective or MOEAs could not generate satisfactory solutions [3, 5, 9]. In that context, a LS technique based on Chung and Church's procedure is applied after each generation, thus orienting the exploration and speeding up the convergence of the EA by refining the chromosomes. Furthermore, two additional mechanisms were incorporated in the evolutionary process in order to avoid the loss of good solutions: an elitism procedure that maintains the best biclusters as well as the diversity in the genotypic space through the generations, and a recovery process that extracts the best solutions of each generation and then copies these individuals into an archive. This archive is actually the set of biclusters returned by the algorithm. Although these two mechanisms appear to be similar to each other, there are several differences between them. The elitism procedure selects the *b* best biclusters that do not overlap in a certain threshold, passing them to the next generation. These solutions can be part of the selection process of further generations allowing production of new solutions based on these by means of the recombination operator. However, due to imperfections on the selection process and of the fitness function, some good solutions can be misplaced through generations. To deal with this issue, we have incorporated an archive, which keeps the best generated biclusters through the entire evolutionary process. It is important to remark that this "meta" population is not part of the selection process, i.e., the evolution of the population after each generation is monitored by the recovery process without interfering in the evolutionary process.

Main Algorithm

As aforementioned, the main loop is a basic evolutionary process that incorporates the LS, the elitism and the recovery procedure. Algorithm 1 illustrates these steps.

Algorithm 1 (Main loop)

Input:	*pop_size*	*(population size)*
	max_gen	*(max number of generations)*
	mut_prob	*(probability of mutation)*
	δ	*(threshold for homogeneity)*
	α	*(parameter for the local search)*
	θ	*(overlap degree of the recovery process)*
	GEDM	*(gene expression data matrix)*
Output:	*arch*	*(a set of biclusters)*

Step 1: *Initialization. Load the data matrix GEDM. Generate a random population P_0 of size pop_size. Generate an empty population arch.*
Step 2: *Main loop. If max_gen is reached, go to Step 9.*
Step 3: *Selection. Perform binary tournament selection over P_t to fill the pool of parents Q_t of size pop_size.*
Step 4: *Elitism procedure. Select at most the best pop_size/2 individuals of P_t that do not overlap each other in at most the 50% of cells. Copy the individuals to P_{t+1}.*
Step 5: *Offspring. Generate the remained (at least pop_size-pop_size/2) individuals of P_{t+1} applying recombination over two random parents of Q_t. Apply uniform mutation to those individuals.*
Step 6: *Local Search. Apply the local search optimization to the individuals of P_{t+1} with mean squared residue above δ.*
Step 7: *Recovery procedure. For each individual $I \in P_{t+1}$ with mean squared residue bellow δ, try to add I to arch in the following way: find the individual $J \in$ arch who shares at least the θ% of cells and then replace J with I only if I is larger than J. If no J where found, add I to arch in an empty slot only if the size of arch is bellow to pop_size. Otherwise discard I.*
Step 8: *End of the loop. Go to Step 2.*
Step 9: *Result. Return arch.*

At this point, the differences between the elitism and the recovery procedure should be clear. The threshold for the overlap level in the elitism procedure, as well as the proportion of elitism, was empirically determined after several runs of the algorithms over different datasets. It is important to note that, as a consequence of a careful design of the recovery procedure, and by means of choosing an adequate value for the θ parameter, the resulting set of biclusters is slightly overlapped in comparison to the high overlapping degree present in the other EAs for biclustering [3, 5, 9].

Individual's Representation
Each individual represents one bicluster, which is encoded by a fixed size binary string built by appending a bit string for genes with another one for conditions. The individual constitutes a solution for the problem of optimal bicluster generation. If a string position is set to 1 the relative row or column belongs to the encoded bicluster, otherwise it does not. Figure 1 shows an example of such encoding for a random individual.

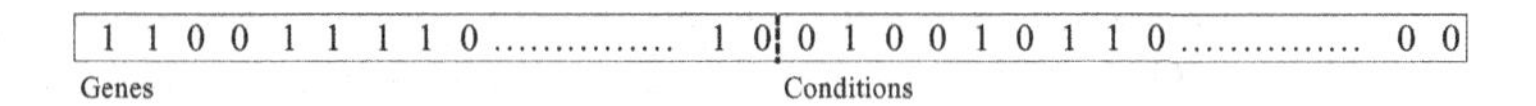

Fig. 1. An encoded individual representing a bicluster

Genetic Operators
After some preliminary tests we decided to apply independent bit mutation to both strings with mutation rates that allow the expected number of bits to be flipped to be the same for both strings. A two point crossover is preferred to one point crossover because the latter would prohibit certain combinations of bits to be crossed over together, especially in cases where the differences in size of rows and columns are notable. In this context, one random point is selected on the rows and the other random point is select over the columns, thus performing the recombination over both search spaces. Then, when both children are obtained combining each one of the two parents' parts, the individual selected to be the descendant is the best in terms of the fitness function.

Fitness Function
As regards the objectives to be optimized, we observed that it was necessary to generate maximal sets of genes and conditions while maintaining the "homogeneity" of the bicluster with a relatively high row variance, as it was established in the equations 1-3. These bicluster features, conflicting to each other, are well suited for multi-objective modeling. An aggregative fitness function that incorporates these features is presented in equation 8. In view of the fact that the local search procedure guarantees the residue constraint [2], the main reason for having a special consideration of the individuals with residue above δ is in the first generation, where the individuals in the population are randomly created. In this context, only those solutions with small residue are preferred. It is also important to consider that an individual can violate the residue constraint during the creation of offspring solutions. Therefore, as the crossover operator returns the best of both children's, individuals with small residue are again preferred in this case, as it can be seen in the fitness function formulation (eq. 8).

$$fitness\ (G,C)=\begin{cases} h(G,C) & if \quad h(G,C)>\delta \\ 1-\dfrac{|G||C|}{mn}+\dfrac{h(G,C)}{\delta}+\dfrac{1}{k(G,C)} & if \quad h(G,C)<=\delta \wedge k(G,C)>1 \\ 1-\dfrac{|G||C|}{mn}+\dfrac{h(G,C)}{\delta}+1 & if \quad otherwise \end{cases} \tag{8}$$

However, when the individuals meet the homogeneity constraint, the LS is not applied. Thus, the improvement of the solutions only depends on the evolutionary process, and then the consideration of biclusters' features such as size, mean squared residue and variance become important. The practical advantage on the consideration of the variance of a bicluster is to avoid constant biclusters [1], since they can be trivially obtained [2]. Note that the fitness function is minimized.

Local Search

The LS procedure that hybridizes the EA was already described. As aforementioned, the greedy approach is based on Chung and Church's work [2], with a small change that avoids the consideration of inverted rows, as applied in [5]. The algorithm starts from a given bicluster (G, C). The genes or conditions having mean squared residue above (or below) a certain threshold are selectively eliminated (or added) according to the description given in the previous sections.

5 Experimental Framework and Results

Two different goals were established for our study. First we need to analyze the quality of the results of BiHEA in the extraction of biclusters with coherent values that follow an additive model. For this analysis, the new approach was tested over synthetic data matrices with different degrees of overlap and noise and then, the results were compared with several of the most important methods for biclustering. Although performing over synthetic data can give an accurate view of the quality of the method, since the optimal biclusters are known beforehand, any artificial scenario inevitably is biased regarding the underlying model and only reflects certain aspects of biological reality. To this end, and in a second experimental phase, we will analyze the biological relevance of the results of BiHEA over a real life data matrix.

Performance Assessment

In order to assess the performance of the biclustering approach over synthetic data, the general bicluster match score is introduced, which is based on the gene match score proposed by [7]. Let M_1, M_2 be two sets of biclusters. The bicluster match score of M_1 with respect to M_2 is given by the equation 9, which reflects the average of the maximum match scores for all biclusters in M_1 with respect to the biclusters in M_2.

$$S^*(M_1, M_2) = \frac{1}{|M_1|} \sum_{(G_1,C_1)\in M_1} \max_{(G_2,C_2)\in M_2} \frac{|(G_1,C_1)\cap(G_2,C_2)|}{|(G_1,C_1)\cup(G_2,C_2)|}. \qquad (9)$$

In this case, instead of considering only the genes of the biclusters of each set [7], the conditions will also be taken into account, i.e., the amount of cells of each bicluster will be assessed. Thus, this measure is more accurate than the metric presented in [7]. Now, let M_{opt} be the set of implanted biclusters and M the output of a biclustering method. The average bicluster precision is defined as $S^*(M, M_{opt})$ and reflects to what extent the generated biclusters represent true biclusters. In contrast, the average bicluster coverage, given by $S^*(M_{opt}, M)$, quantifies how well each of the true biclusters is recovered by the biclustering algorithm under consideration. Both scores take the maximum value of 1 if $M_{opt}= M$.

As regard the real data, since the optimal biclusters are unknown, the above metric can not be applied. However, prior biological knowledge in the form of natural language descriptions of functions and processes to which the genes are related has become widely available. Similar to the idea pursued in [7, 11, 12], whether the groups of genes delivered by BiHEA show significant enrichment with respect to a specific Gene Ontology (GO) annotation will be investigated. Then, a novel measure was designed in order to assess the molecular function and biological process enrichment of the results of a biclustering method. Let M be a set of biclusters, GO a specific GO annotation and α a statistic significant level. The overall enrichment indicator of M with respect of GO on a statistically significant level of α is given by:

$$E^{*}(M,GO,\alpha)=\frac{1}{|M|}\sum_{(G,C)\in M}\frac{Maxenrichment(G,GO,\alpha)}{|G|}\sum_{(G,C)\in M}\frac{|G|}{n}. \qquad (10)$$

where $Maxenrichment(G, GO, \alpha)$ is the maximum gene amount of G with a common molecular function/biological process under GO with a statistically significant α level. The metric of equation 10 measures the average of the maximum gene proportion statistically significant of molecular function/biological process enrichment of a set of biclusters M on a specific GO annotation, pondered with the average genes of M. It is an indicator of the quality of the results of a clustering/biclustering method on real data, and can be used to compare several methods, being the highest values the best.

First Experimental Phase: Synthetic Data

Data preparation

The artificial model used to generate synthetic gene expression data is similar to the approaches proposed by [7, 13]. In this regard, the biclusters represent transcription modules, where these modules are defined by a set G of genes regulated by a set of common transcription factors and a set C of conditions in which these transcription factors are active. Varying the amount of genes and conditions that two modules have in common, it is possible to vary the overlap degree in the implanted biclusters. To this end, we define the overlap degree d, as an *indicator* of the maximum amount of cells that two transcription modules can share. The amount of shared cells is actually d^2.

This model enables the investigation of the capability of a method to recover known groupings, while at the same time, further aspects like noise and regulatory complexity can be systematically studied [7]. The datasets are kept small, n = 100 and m = 100. This, however, does not restrict the generality of the results. In the case of d = 0, 10 non-overlapped biclusters (size = 10 rows *times* 10 columns) were implanted. For every d>0, the size of the artificial biclusters was increased in d rows and d columns, except for the rightmost bicluster, for which its size remains unchanged. For d>1, 18 additional biclusters appear in the data matrices, as a consequence of the overlap of the implanted transcription modules. These extra biclusters are also included in our study since they are equally suitable for being extracted by a biclustering method, although the overlap degree is higher than for the artificial transcription factors. The figure 2 depicts the previous scenario. The values of each bicluster were determined as follows: for the first row, random real numbers between 0 and 300 from a uniform distribution were incorporated. Then, for each one of the remainder rows, a unique random value between 0 and 300 is obtained and added to each element of the first row. The result is a bicluster with coherent values under an additive model [1], with a mean squared residue equal to 0 and a row variance greater than 0. The remainder slots in the matrix were filled with random real values between 0 and 600.

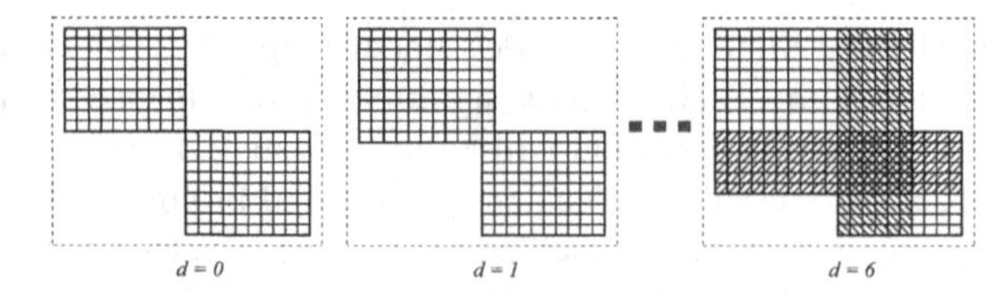

Fig. 2. Overlapping levels of artificial biclusters according to *d*. In *d*=6, the diagonal lines represent the extra biclusters generated by the overlapping of the implanted biclusters.

Synthetic datasets built following the aforementioned procedure are useful to analyze the behavior of a biclustering method in increasing regulatory complexity levels. However, these datasets represent an ideal scenario without noise, i.e., far away from realistic data. To deal with this issue, and in view of the fact that real scenarios have a great regulatory complexity, the behavior of this proposal with *d*=6 and with increasing noise levels will be also investigated. For the noisy model, the expression values of the biclusters are changed adding a random value from a uniform distribution between -*k* and *k*, with *k*=0, 5, 10, 15, 20 and 25 to each cell.

Results

For referential purposes, several important biclustering algorithms were run: BiMax, CC, OPSM, ISA, and SPEA2^{LS}. For the first four implementations, the BicAT [14] tool was used. All the parameters for these methods were set after several runs, in order to obtain the best results of each strategy. For BiHEA, the parameters' setting is the following: population = 200; generations = 100; δ = 300; α = 1.2; mutation probability = 0.3; and θ = 70. Since the number of generated biclusters strongly varies among the considered methods, a filtering procedure, similar to the recovery process of our approach, has been applied to the output of the algorithms to provide a common basis for a fair comparison. The filtering procedure extracts, for each of the resulting set of biclusters, at most *q* of the largest biclusters that overlap in at most the θ=70% of cells. For $d<2$, *q* is set to 10, and for the rest, *q* is set to 28. As regards the results, in figures 3a and 3b, the average precision and the average coverage obtained by the different biclustering methods are shown, for the scenarios with increasing overlapping degrees. Similarly, in figures 3c and 3d, the results for the scenarios with increasing noise levels are illustrated.

As it can be observed, BiHEA outperforms the referential methods in all the scenarios, in terms of both the precision and the coverage of biclusters. As the overlapping degree increases, figures 3a and 3b show that the results obtained by our method improve, reaching an almost perfect score with *d*=6. This can be explained in terms of the theory of basic schemes in genetic algorithms, since in higher degrees of overlap, useful schemes shared between the optimal biclusters are larger in size. This feature facilitates the construction of solutions that meet the homogeneity constraint by means of the crossover operator. The last observation should be true for most EAs. Nonetheless, the imperfections on the selection process and fitness functions can derive on a misuse of this advantage, as it happens with SPEA2^{LS}. This clearly shows the need of the recovery process introduced on the BiHEA.

In regard to the effects of noise, the results are the expected ones. As the levels of noise augment, the degradation of a perfect bicluster increases the residue and

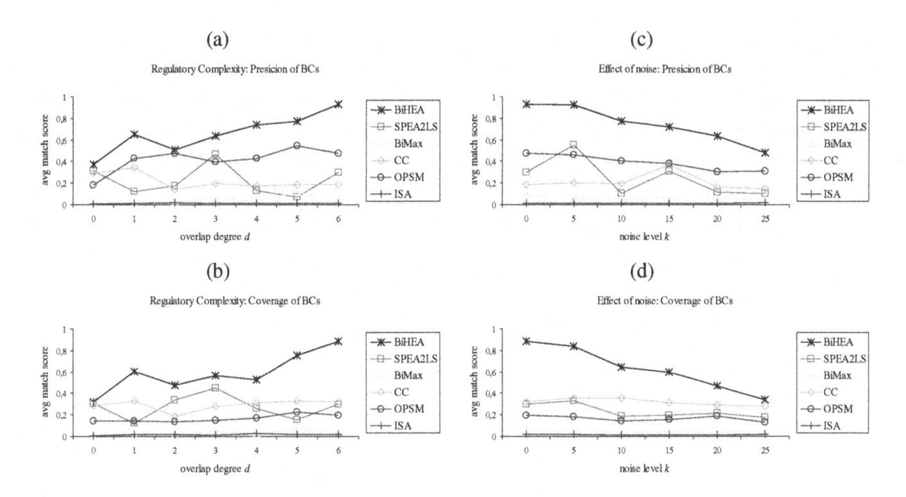

Fig. 3. Results for the artificial scenarios. Figures 3a and 3b show the average precision and the average coverage respectively in overlapped scenarios. Figures 3c and 3d show the average precision and the average coverage respectively in noisy scenarios.

possibly, the homogeneity constraint can no longer be satisfied for the entire bicluster. For the reference methods, OPSM, CC and SPEA2^{LS} show similar results, OPSM being the more precise one although the coverage appears to be worse than the results achieved by CC and SPEA2^{LS}. However, these methods appear to be less susceptible to the noise than BiHEA. On the other hand, both BiMax and ISA can not obtain significant biclusters, which contrasts with the conclusions published in [7] where both methods achieve almost perfect scores. Nevertheless, we argue that this may be a consequence of the way in which synthetic data are constructed, since in the case of BiMax, the discretization method is unable to obtain an appropriate binary representation of the synthetic data matrices. On the other hand, the notion of similarity of rows and columns in the ISA algorithm might be different from the one used here. However, the synthetic data used in this work was designed in the aforementioned manner since, according to our knowledge it represents general and relevant GEDMs, which allow a fair comparison in the evaluation of the algorithms.

Second Experimental Phase: Real Data

In this subsection, the results of BiHEA on a real GEDM will be briefly analyzed. This study will be focused on a colon cancer data [15] that consists in a GEDM of 62 colon tissue samples, 22 of which are normal and 40 are tumor tissues. This analysis will be focused on the 2000 genes with the highest minimal intensity [15]. For the experimentation, an ontological analysis of the 10 first resulting biclusters found by BiHEA, CC, ISA, OPSM and SPEA2^{LS} will be performed. The BiMax algorithm is not included since an adequate parameter setup for the discretization step could not be found. The parameters for the proposed approach remain almost the same, except for the following: $\delta = 150$; $\alpha = 2.0$. All the ontological classification was performed with the ontology tool Onto-Express [12], applying a hyper geometric distribution and referencing the calculations by the 2000 genes analyzed.

As regard the results, the figure 4 depicts the values achieved by the previous methods in terms of the overall enrichment indicator (cf. eq. 10 with $\alpha = 0.05$) for molecular function and biological process enrichment. It is clear that BiHEA is the method that obtains the better results, since the quality of the outcomes outperforms the results of the referential algorithms in terms of the overall enrichment indicator. Only OPSM remains close, whereas the other approaches obtain significantly worse results. These results are consistent with the ones obtained on the synthetic datasets, thus showing the correctness of the artificial model selected.

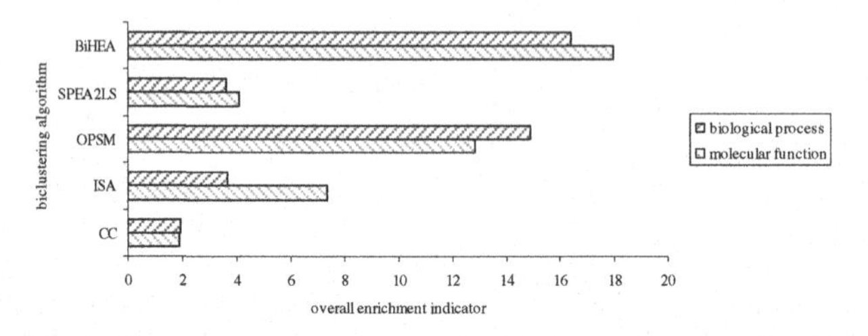

Fig. 4. Overall enrichment indicator of BiHEA, SPEA2^{LS}, OPSM, ISA and CC for molecular function and biological process enrichment

6 Conclusions

In this paper, we have introduced a new memetic evolutionary approach for microarray biclustering. The original EA was hybridized with a LS procedure for finer tuning, and also two novel features were introduced: the first one was designed in order to avoid the loss of good solutions through generations, while keeping a low degree of overlap between the final biclusters, and the other one was conceived so as to maintain a satisfactory level of diversity in the genotypic space.

In a first experimental phase on synthetic datasets, the results obtained with our method outperform the outcomes of several biclustering approaches of the literature, especially in the case of coherent biclusters with high overlap degrees. Nonetheless, this can not be considered as a drawback because, in general, the regulatory complexity of an organism is far from the model of non-overlapped biclusters. Furthermore, an analysis on a real dataset was performed and, in terms of the proposed measure, the quality of the outcomes of BiHEA is clearly better than the results of the reference methods. In fact, this shows the correctness of the model designed to build the biclusters, i.e., coherent biclusters following an additive model. Although this is consistent with the results obtained in the synthetic datasets, an extensive analysis on several real datasets in needed to confirm these results.

Finally, the framework for the comparison of biclustering algorithms was refined by means of the introduction of two new measures: the bicluster match score S^* and the overall enrichment indicator E^*. The first one is useful to test on synthetic data since the optimal biclusters are known beforehand. The last one can be used to assess the performance of several methods in real data in terms of a specific GO annotation. Both metrics are indispensable in any quality assessment of biclustering algorithms since they provide a fair framework in which the methods can be compared.

Acknowledgements. Authors acknowledge CONICET, the ANPCyT for Grant N°11-1652 and SeCyT (UNS) for Grant PGI 24/ZN15.

References

1. Madeira, S., Oliveira, A.L.: Biclustering Algorithms for Biological Data Analysis: A Survey. IEEE-ACM Trans. Comput. Biol. Bioinform. 1, 24–45 (2004)
2. Cheng, Y., Church, G.M.: Biclustering of Expression Data. In: Proceedings of the 8th Inter-national Conf. on Intelligent Systems for Molecular Biology, pp. 93–103 (2000)
3. Mitra, S., Banka, H.: Multi-objective evolutionary biclustering of gene expression data. Pattern Recognit. 39, 2464–2477 (2006)
4. Divina, F., Aguilar-Ruiz, J.S.: Biclustering of Expression Data with Evolutionary Computation. IEEE Trans. Knowl. Data Eng. 18(5), 590–602 (2006)
5. Bleuler, S., Prelic, A., Zitzler, E.: An EA framework for biclustering of gene expression data. In: Proceeding of Congress on Evolutionary Computation, pp. 166–173 (2004)
6. Ihmels, J., Bergmann, S., Barkai, N.: Defining transcription modules using large-scale gene expression data. Bioinformatics 20(13), 1993–2003 (2004)
7. Zimmermann, P., Wille, A., Buhlmann, P., Gruissem, W., Hennig, L., Thiele, L., Zitzler, E., Prelic, A., Bleuler, S.: A systematic comparison and evaluation of biclustering methods for gene expression data. Bioinformatics 22(9), 1122–1129 (2006)
8. Ben-Dor, A., Chor, B., Karp, R., Yakhini, Z.: Discovering Local Structure in Gene Expression Data: The Order-Preserving Submatrix Problem. In: Proc. Sixth Int'l Conf. Computational Biology (RECOMB 2002), pp. 49–57 (2002)
9. Gallo, C., Carballido, J.A., Ponzoni, I.: Microarray Biclustering: A Novel Memetic Approach Based on the PISA Platform. LNCS, vol. 5483, pp. 44–55. Springer, Heidelberg (2009)
10. Zitzler, E., Laumanns, M., Thiele, L.: SPEA2: Improving the strength pareto evolutionary algorithm for multiobjective optimization. In: Giannakoglou, Tsahalis, Periaux, Papailiou, Fogarty (eds.) Evolutionary Methods for Design, Optimisations and Control, pp. 19–26 (2002)
11. Tanay, A., et al.: Discovering statistically significant biclusters in gene expression data. Bioinformatics 18(suppl. 1), S136–S144 (2002)
12. Draghici, S., Khatri, P., Bhavsar, P., Shah, A., Krawetz, S., Tainsky, M.: Onto-Tools, the toolkit of the modern biologist: Onto-Express, Onto-Compare, Onto-Design, and Onto-Translate. Nuc. Acids Res. 31(13), 3775–3781 (2003)
13. Ihmels, J., et al.: Revealing modular organization in the yeast transcriptional network. Nat. Genet. 31, 370–377 (2002)
14. Barkow, S., Bleuler, S., Prelic, A., Zimmermann, P., Zitzler, E.: BicAT: a biclustering analysis toolbox. Bioinformatics 22(10), 1282–1283 (2006)
15. Alon, U., Barkai, N., Notterman, D., Gish, K., Ybarra, S., Mack, D., Levine, A.: Broad patterns of gene expression revealed by clustering analysis of tumor and normal colon tissues probed by oligonucleotide arrays. Proc. Natl. Acad. Sci. 96, 6745–6750 (1999)

Using Supervised Complexity Measures in the Analysis of Cancer Gene Expression Data Sets

Ivan G. Costa[1], Ana C. Lorena[2],
Liciana R.M.P. y Peres[2], and Marcilio C.P. de Souto[3]

[1] Center of Informatics, Federal University of Pernambuco, Recife, Brazil
igcf@cin.ufpe.br
[2] Center of Mathematics, Computation and Cognition, ABC Fed. Univ., SP, Brazil
ana.lorena@ufabc.edu.br, liciana.perez@ufabc.edu.br
[3] Dept. of Informatics and Applied Mathematics, Fed. Univ. of Rio Grande do Norte
marcilio@dimap.ufrn.br

Abstract. Supervised Machine Learning methods have been successfully applied for performing gene expression based cancer diagnosis. Characteristics intrinsic to cancer gene expression data sets, such as high dimensionality, low number of samples and presence of noise makes the classification task very difficult. Furthermore, limitations in the classifier performance may often be attributed to characteristics intrinsic to a particular data set.

This paper presents an analysis of gene expression data sets for cancer diagnosis using classification complexity measures. Such measures consider data geometry, distribution and linear separability as indications of complexity of the classification task. The results obtained indicate that the cancer data sets investigated are formed by mostly linearly separable non-overlapping classes, supporting the good predictive performance of robust linear classifiers, such as SVMs, on the given data sets. Furthermore, we found two complexity indices, which were good indicators for the difficulty of gene expression based cancer diagnosis.

Keywords: Cancer gene expression classification, Machine Learning, data set complexity.

1 Introduction

Technologies for measuring the gene expression of complete cell genomes have paved the way towards personalized medicine [29]. In other words, diagnoses of diseases can be based on molecular level information of individual patients, which enhances the accuracy of diagnoses in relation to classical methods. In this context, supervised Machine Learning (ML) methods have been successfully applied for performing gene expression based cancer diagnosis [27]. However, characteristics intrinsic to cancer gene expression data sets, such as high dimensionality, low number of samples and presence of noise could make the classification task performed by ML methods harder [13,27].

K.S. Guimarães, A. Panchenko, T.M. Przytycka (Eds.): BSB 2009, LNBI 5676, pp. 48–59, 2009.

A great deal of research in supervised ML has focused on the development of algorithms able to build competitive classifiers in terms of generalization ability and computational time. Classification using ML techniques consists of inducing a function $f(\mathbf{x})$ from a known training data set composed of n pairs $(\mathbf{x}_i, y_i)$, where $\mathbf{x}_i$ is an input data and y_i corresponds to its class [18]. As previously mentioned, the induced function (classifier) should be able to predict the class of new data for which the classification is unknown, performing the desired discrimination (generalization ability). However, as stated in [12,2], limitations in the classifier performance may be often attributed to difficulties intrinsic to the data set. In such a context, data set complexity analysis is a recent area of research that tries to characterize the intrinsic complexity of a data set and find relationship with the accuracy of classifiers.

In this paper, we analyze this issue of classification complexity in the context of gene expression based cancer diagnostic. In a previous work [15], we showed that simple, often linear classifiers, have a good accuracy performance in the discrimination of the cancer classes, indicating that they are mainly linearly separable. As stated before, gene expression data sets are often characterized by a high dimensionality, while presenting a small number of examples. The fact these data sets are often linear separable could be explained by the curse of dimensionality: the few samples in the data set are sparsely distributed in the feature space, thus there is a high probability of finding a hyperplane separating data points of distinct classes [11]. Even though this is a well-known fact in the cancer gene expression literature [4], data set complexity and linear separability has not been consistently addressed before.

This work complements the analysis of cancer gene expression data sets in [15] by the use of complexity indices presented in [12]. These indices characterize the difficulty of a classification problem, focusing on the complexity of the class boundary. They measure aspects such as topological and geometrical characteristics of the class distribution. Furthermore, some indices are based on well-established procedures to identify problems that are linearly separable — linearly separable problems are considered the easiest ones in classification tasks. These complexity indices are calculated for 10 gene expression data sets. We use these data sets as inputs to build classifiers using Naïve-Bayes, logistic Regression, linear Support Vector Machines (SVM), k-nearest neighbors (KNN) and voted Perceptron. Next, we analyze the distribution of complexity indices of the data sets, and the relation of the indices and the error rate of the classification methods.

This paper is structured as follows: Section 2 describes gene expression data analysis for cancer classification. Section 3 presents the materials and methods employed in the experiments performed in this work. Section 4 presents the experimental results of this work, while Section 5 presents the main conclusions of the paper.

2 Classification of Gene Expression Data

Genes are portions of DNA molecules present in the cells [1]. They store the organisms' genetic information. For instance, genes codify proteins, the main building blocks of all organisms, which are synthesized in a process called gene expression. Indeed, the expression level of a gene constitutes an estimate of the amount of proteins it produces in a given period.

One of the main techniques employed to measure the expression levels of genes is the microarray technology, which allows to quantify the expression level of thousands of genes simultaneously [21]. Several different kinds of experiments can be performed using microarrays. For example, they can be performed for comparing gene expression levels in different types of tissues (e.g., normal and tumor tissues). The latter can be used in the diagnosis of diseases, through the classification of different types or subtypes of tumors according to their expression patterns [5,6,28,33]. Other possible analysis is the identification of genes that are most related to a particular disease, as presented in the seminal work in [10], which could then be target for future medicines and genetic therapies.

Our work is mainly concerned with the analysis of the complexity of gene expression data related to cancer diagnosis [5,6,28,32,33].

Cancer diagnosis is generally performed by a series of microscopic and immunologic tissue tests. Often, the presence of tumor samples with atypical morphologies makes these analysis difficult [22]. Furthermore, some cancer tissues from different types of tumors (or subtypes) show low differentiation. This makes laboratorial identification based only on morphology and immunophenotyping difficult. In this scenario, microarray could be used to characterize the molecular variations among tissues by monitoring gene expression profiles on a genomic scale [10,5,21,33].

In fact, the analysis presented here is an extension of our previous work [15], where we studied the complexity of nine gene expression data related to cancer diagnosis, measured using cDNA plataform (double-chanel). In this current investigation, we analyze not only microarray data collected via cDNA platform, but also with Affymetrix (single-chanel). Furthermore, instead of using only the performance of a set of classifiers to draw conclusions, we employ various data complexity measures proposed in [12]. In this context, the work mostly related to ours is the one in [20]. But, differently from our work, whose aim is to present an extensive study of the complexity of different data sets, the main goal in [20] was the proposal of a scheme to build ensemble of classifiers (k-Nearest Neighbor as base classifier) employing the data set complexity measures as guide.

3 Materials and Methods

In this section, we present the main aspects taken into account in the analysis performed. We first describe the data sets used in the experiments and, then, discuss the empirical aspects considered in the set of experiments carried out for this work.

3.1 Data Sets

Ten microarray data sets are included in the experiments (Table 1). They are a set of benchmark microarray data first presented in [3][1]. All of them have two

[1] The data sets analyzed in this study are available at http://algorithmics.molgen.mpg.de/Supplements/CompCancer/

Table 1. Data sets description

Data set	Tissue	Chip	n	Dist. Classes	d	Filtered d
Alizadeh-2000-v1	Blood	cDNA	42	21,21	4022	1095
Armstrong-2002-v1	Blood	Affy	72	24,48	12582	1081
Bittner-2000	Skin	cDNA	38	19,19	8067	2201
Chowdary-2006	Breast, Colon	Affy	104	62,42	22283	182
Gordon-2002	Lung	Affy	181	31,150	12533	1626
Laiho-2007	Colon	Affy	37	8,29	22883	2202
Pomeroy-2002-v2	Brain	Affy	34	25,9	7129	857
Shipp-2002-v1	Blood	Affy	77	58,19	7129	798
West-2001	Breast	Affy	49	25,24	7129	1198
Yeoh-2002-v1	Bone marrow	Affy	248	43,205	12625	2526

classes, representing different subtypes of tumors. As Table 1 illustrates, these data sets present different values for aspects like type of tissue (second column), type of array chip (third column), number of data items (fourth column), distribution of data within the classes (fifth column), dimensionality (sixth column), and dimensionality after feature selection (last column). The experiments will be developed with the data sets after the feature selection (last column). This feature selection was based on discarding genes with low variance, where the variance threshold was selected to keep around 10% of genes (see [3] for details).

In terms of the data sets, microarray technology is generally available in two different types of platforms, single-channel microarrays (e.g., Affymetrix) or double-channel microarrays (e.g., DNA) [19,21,24]. Other microarrays technologies are also based on either single and double channels methods. As the data sets analyzed here are restricted to those collected with cDNA and Affymetrix microarrays, we employ the terms cDNA and Affymetrix to denote double or single-channel arrays, respectively.

3.2 Data Complexity Measures

As discussed in [12], a classification problem can be difficult for different reasons. For example, the classes can be ambiguous either intrinsically or for inadequate feature (attribute) measurements. Other problems could present a complex decision boundary and/or subclass structure in such a way that no compact description of the boundary is possible. These can hold true independently of sample size (training set) or feature space dimensionality (the size of the vector representing the input data). Also, small sample size and high dimensionality induce often sparsity, introducing another layer of difficulty via a lack of constraints on the generalization rules for the classifier built. Indeed, if the sample (training set) is too sparse, an intrinsically complex problem could appear deceptively simple.

In order to measure these aspects, [12] proposed a set of indices. Such indices are, by definition, only suitable for data sets presenting two classes. These complexity indices falls into three main classes: measures of overlap, measures of linear separability and measures of topology. As indices of each class are correlated, we implement only a subset of the indices. Before presenting the definitions for these indices, we will introduce some basic definitions needed to understand them. Let X be a d by n matrix representing a gene expression data set, where x_{ij} denotes the expression value of sample j and feature i, x_i is a d-dimensional vector with the expression values of sample (patient) i and $x_{.j}$ is a n-dimensional vector representing the expression values of feature (gene) j.

Fisher's Discrimination Ratio ($F1$). This index calculates the discriminative power of a feature. In the case of multidimensional variables, the value of the variable with maximum discrimination ratio is chosen, as only one feature is enough to discriminate the classes.

$$F1 = \max_j \frac{(\mu_{1j} - \mu_{2j})}{\sigma_{1j}^2 + \sigma_{2j}^2} \tag{1}$$

where μ_{1j} and μ_{2j} are the mean of the two classes for feature j; σ_{1j}^2 and σ_{2j}^2 are the variances of the two classes for feature j.

Volume of Overlapping Region ($F2$). This index measures the length of the overlap between the distributions of values of the two classes. The overlap size is normalized by the total length of the distribution of both classes. Let $max_1(x_{.j})$ and $min_1(x_{.j})$ be the maximum and minimum value of feature i at class 1.

$$F2 = \prod_i \frac{min(max_1(x_{.j}), max_2(x_{.j})) - max(min_1(x_{.j}), min_2(x_{.j}))}{max(max_1(x_{.j}), max_2(x_{.j})) - min(min_1(x_{.j}), min_2(x_{.j}))} \tag{2}$$

Linear Separability Indices($L1$ and $L2$). As in [12], we use a linear programming method for performing linear classification of the classes. This is done by the following formulation [25]:

$$\begin{aligned} \texttt{minimize} \quad & a^t t \\ \texttt{subjectto} \; & Y^t w + t \geq b \\ & t \geq 0 \end{aligned}$$

where a and b are vectors with entries equal to 1, w is the weight vector, t is the error vector and Y is a matrix representing the expression data, such that $y_i = x_i$ for samples of class 1 and $y_i = -x_i$ for samples of class 2.

$L1$ is obtained by summing the vector t, i.e., the sum of the distance of samples to the linear boundary. Another measure ($L2$) is achieved by estimating the classification error rate of the linear classifier. Both measures give an estimate on how linearly separable the classes are.

Mixture Identifiability Index ($N1$). The index $N1$ measures if the two classes comes from distinct distributions. Given a similarity matrix of samples, estimate the minimum spanning tree and count the proportion of edges connecting samples from distinct classes [8].

Nearest Neighbors Indices ($N2$ and $N3$). $N2$ is based on measuring for each sample the distance of a nearest neighbor in the same class and in the distinct class. Then, the ratio of the average of all inter and intra class distances is calculated. This index measures how disperse are intra class samples and how close are inter class samples. $N3$ is calculated by the classification error of a leave-one-out nearest neighbor classifier. In all complexity measures based on distances/similarity ($N1, N2, N3$), euclidian distance was used.

Dimensionality/Samples Ratio ($T1$). This index is based on the log of the ratio of number of features against number of samples: $T1 = \log(d/n)$.

3.3 Classification Algorithms

This section describes the classification algorithms employed in the experiments. Simple classification techniques, mainly generating linear classification boundaries, were chosen. The objective was to evaluate whether the performance of these simple classification techniques could be considered satisfactory, confirming the fact that gene expression data sets for cancer diagnosis have a simple classification structure.

Naïve-Bayes. Naïve Bayes (NB) are probabilistic classifiers based on the Bayes theorem for conditional probabilities. They build a function, to be optimized, using a constrained (naïve) assumption that all attributes in a data set are independent, given the class. Therefore, NB assumes that the presence/absence of a gene describing a certain class is unrelated to the presence/absence of any other gene, which usually does not hold. NB training is in general performed through the use of maximum likelihood algorithms. Despite its simplicity, NB have been successful in complex practical applications, specially in text mining [17]. It also shows low train and prediction times.

Logistic Regression. Logistic Regression (LR) classifiers are statistical models in which a logistic curve is fitted to the data set [14], modeling the probability of occurrence of a certain class. LR classifiers are also known as: logistic models, logit models or maximum-entropy classifiers. The first step in LR consists of building a logit variable, containing the natural log of the odds of the class occurring or not. A maximum likelihood estimation algorithm is then applied to estimate the probabilities. LR models are largely employed in Statistics and have demonstrated success in several real-world problems.

Support Vector Machines. Support Vector Machines (SVMs) are based on concepts from the Statistical Learning Theory [30]. In its linear form, which was employed in this work, given a dataset T composed of n pairs $(\mathbf{x}_i, y_i)$, in which $\mathbf{x}_{.j} \in \Re^m$ and $y_i \in \{-1, +1\}$, SVMs seek for a hyperplane $\mathbf{w} \cdot \mathbf{x} + b = 0$ able to separate the data in T with minimum error while maximizing the margin of separation between the classes. SVMs have good generalization ability. Besides, SVMs also stand out for their robustness to high dimensional data. Their main deficiency concerns the difficulty of interpreting the generated model and their sensibility to a proper parameter tuning.

Voted Perceptron. One of the first algorithms proposed for the linear separation of a data set is the *Perceptron* Neural Network [23]. It performs incremental changes in the parameters $(\mathbf{w}, b)$ of a linear function until all data are correctly classified. In [7], a generalization of the Perceptron algorithm named *"Voted-Perceptron"* (VP) was proposed. VP maintains all vectors $\mathbf{w}_i$ generated during Perceptron training. For each vector $\mathbf{w}_i$, a counter c_i is also kept, which stores the number of iterations for which the vector stands until an error is committed. This number defines a weight attributed to the function correspondent to $\mathbf{w}_i$. The final prediction in VP is then given by:

$$g(\mathbf{x}) = \operatorname{sgn}\left(\sum_{i=1}^{k} c_i \operatorname{sgn}(\mathbf{w}_i \cdot \mathbf{x})\right) \tag{3}$$

The rule presented in Equation 3 represents a majority voting of the classification outputs of several linear decision functions. Therefore, VP can be visualized as an ensemble technique. Experiments in [7] demonstrated that VP had an accuracy performance superior to the traditional Perceptron, although it was inferior to that of SVMs.

k-Nearest Neighbor. The k-Nearest Neighbor (KNN) algorithm is a simple representative of instance-based ML techniques. It stores all training data and classifies a new data point according to the class of the majority of its k nearest neighbors in the training data set. To obtain the nearest neighbors for each data, KNN computes the distance between pairs of data items, usually using the Euclidean distance. A great advantage of KNN is its simplicity. Nevertheless, KNN prediction times are usually costly, since all training data may need to be revisited.

4 Experiments and Results

First, we evaluated, using leave-one-out, the error rate of five classification methods: Naïve Bayes (NB), Logistic Regression (LR), Linear SVMs, Voted Perceptron (VP) and k-Nearest Neighbor (KNN) for these data sets. In order to put the results in perspective, we also include in the last column, the error of a classifier that always gives as output the class of largest number of samples (majority class). We will refer to this classier as base classifier (BC). The error rate of each method for each data set is shown in Table 2. For each data set, best results are highlighted in bold-face. All classifiers were induced using the Weka simulator [31], with default parameter values. Some results could not be calculated for the Logistic Regression due to computational constraints. These are indicated with a "-" in the table. Then, we calculate the seven classification complexity measures proposed in [12] for each of the 10 cancer gene expression data sets.

In terms of the performance of the classifiers generated, one can observe from Table 2 that, in general, all classifiers errors rates were smaller than the one produced by the base classifier (BC) for all data sets, indicating that they were able to generalize for the given data sets. More specifically, linear SVMs presented

Table 2. Classification error rates of distinct methods

	(NB)	(LR)	SVM	VP	KNN	BC
Alizadeh-2000-v1	**4.76**	7.14	7.14	14.29	23.81	50.00
Armstrong-2002-v1	**1.39**	5.56	2.78	8.33	9.72	33.33
Bittner-2000	21.05	-	**13.16**	21.05	28.95	50.00
Chowdary-2006	**2.88**	4.81	3.85	4.81	3.85	40.39
Gordon-2002	**0.55**	**0.55**	**0.55**	2.76	1.10	17.03
Laiho-2007	21.62	-	**2.70**	32.43	8.11	21.62
Pomeroy-2002-v2	20.59	20.59	**17.65**	26.47	32.35	26.47
Shipp-2002-v1	18.18	14.29	**6.49**	18.18	19.48	24.68
West-2001	**14.29**	18.37	16.33	20.41	24.49	48.98
Yeoh-2002-v1	4.03	-	**1.61**	**1.61**	8.47	17.34

often smaller error rates for all data sets: in six out of the 10 data sets analyzed. NB had best performance in five out of the 10 data sets. These results are in accordance to those already presented in [15] and indicate that the investigated data sets show a linear classification structure.

For a further investigation, we drew a scatter plot of all methods versus the complexity measures calculated (Figure 1). Next, in order analyze if the classification error rate of a given method is related to a specific complexity measure, we calculated the correlation coefficient and perform a t-test [26]. The correlation coefficients is written at the upper corner of each plot: bold-faced values indicates statistically significant correlations (p-value < 0.05). In the figure we omit indices $L1$ and $F2$, as their values are all very close to zero and had low correlation coefficients.

Looking at the indices themselves and their distributions, we find some interesting properties of the data sets. $F2$ was equal to zero in nine out of ten data sets. This indicates that there is no overlap between classes for such data sets. $L1$ also have results very close to zero in all data sets ($< 10^{-10}$), which indicates that examples incorrectly classified by a linear classifier are very close to the class boundary. Moreover, $L2$ was equal to zero for six out of the ten data sets, which is strong evidence that most of data sets are linearly separable. All these are likely a consequence of the curse of dimensionality: a problem well-known to be related to cancer gene expression data sets [4,16], but often ignored in many cancer gene expression studies. In such data sets, few samples are sparsely distributed in a large multidimensional space, making classes to be non-overlapping and linearly separable [11]. We also find that indices $N1$, $N2$ and $N3$ are highly positively correlated (p-value < 0.05). This indicates that these indices, which are based on capturing classification complexity in a data distribution perspective, are equivalent in these data sets. Furthermore, $N1$ and $T1$ are negatively correlated (p-value < 0.05), which indicate that complexity measured by $N1$ is at least in part a consequence of the curse of dimensionality.

As illustrated in Figure 1, the error rate of all classifiers generated are negatively correlated to $T1$ (p-value < 0.05). Often, this sort of behavior would be

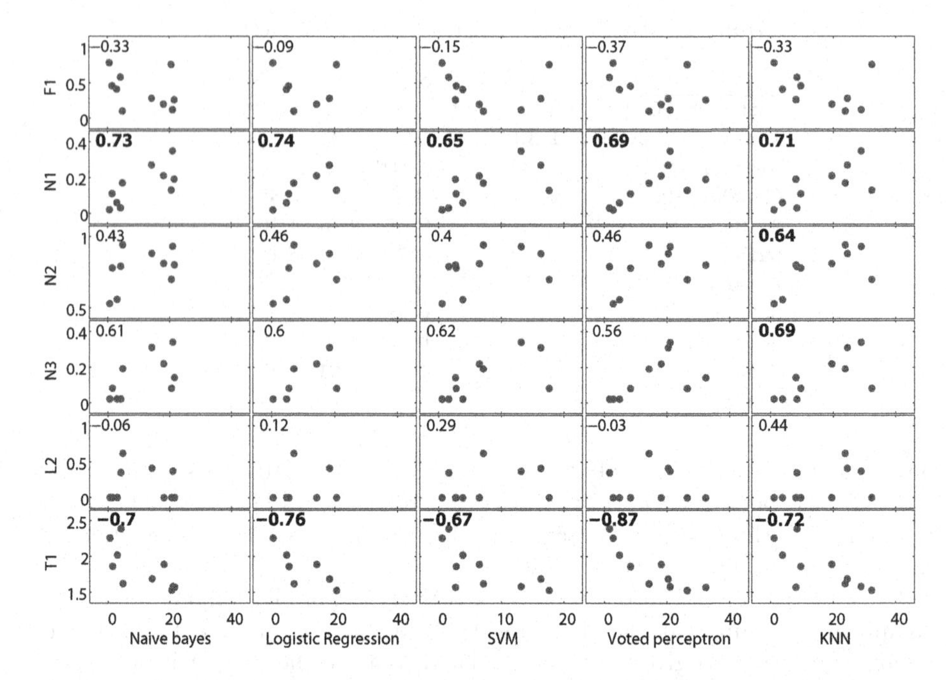

Fig. 1. Scatter plot of complexity indices versus classification accuracies of classification methods

expected as the relation of sample size/dimensionality is a crucial to general classification problems, in particular for cancer gene expression data sets.

The error rate of NB, SVMs, VP and KNN were positively correlated to $N1$. This index measures the degree that class labels forms two distinct distributions, where higher values indicate smaller separation in distributions and a harder classification task. Furthermore, KNN was positively correlated with $N2$ and $N3$, which is expected as these indices are based on rules similar to the KNN classifier. Interestingly, $F1$ has a low/moderate correlation to the classification accuracies of the classifiers, but in most cases the plots have a curve-like distribution. This indicates some non-linear dependency of the variable, which cannot be detected by the correlation test. Possibly, as explored in [12], pair of complexity indices could give better insights towards description of classification complexity. We could not find any interesting correlation using pairs of variables, possibly given the low number of data sets available for such study. Note that we are restricted by the complexity indices to data set with two classes only, and collecting such data sets is a laborious work.

5 Conclusions

We performed here the first study of a set of data complexity measures in the context cancer gene expression data sets. Such an analysis is very important prior to the development of any automatic diagnostic method, as it can indicate, looking only at data set characteristics, if the classification task can be accurately be performed. Our results indicate that the cancer data sets that we investigated are formed by mostly linearly separable non-overlapping classes. This was reinforced by the fact that linear SVMs obtained the best classification accuracy in most data sets. Furthermore, we could also show that two complexity indices correlate with the classification accuracy of the methods: $T1$ and $L2$. The correlation with $T1$ indicates that a low number of patient samples imposes limits in the accuracy of any of the classification methods. $N2$, an index that measures the complexity of distribution of samples, was also a good indicator of hardness of the classification task.

As future work, we intend to investigate the classification complexity of data sets after the application of several gene filters, which are usually applied prior to the classification methods. Furthermore, we will explore the use of complexity measures for meta-learning tasks. In meta-learning, descriptors of data sets are collected and one tries to build a classifier to choose the more appropriate ML method for a given data set [9]. Such methodology would allow the indication of a most suitable classification method for a given data set.

Acknowledgments. We thank the referees for their valuable comments. This work has been partially supported by Brazilian research agencies: FAPESP, FACEPE, CNPq and CAPES.

References

1. Alberts, B., Al, E.: Molecular Biology of the Cell. Garland Science (2002)
2. Bernadó-Mansilla, E., Maciá-Antonilez, N.: Modeling problem transformation based on data complexity. In: Angulo, C., Godo, L. (eds.) Artificial Intelligence Research and Development, pp. 133–139. IOS Press, Amsterdam (2007)
3. de Souto, M.C.P., Costa, I.G., de Araujo, D.S.A., Ludermir, T.B., Schliep, A.: Clustering cancer gene expression data: a comparative study. BMC Bioinformatics 9, 497+ (2008)
4. Dudoit, S., Fridlyand, J., Speed, T.P.: Comparison of discrimination methods for the classification of tumors using gene expression data. Journal of the American Statistical Association 97(457), 77–87 (2002)
5. Dudoit, S., Fridlyand, J., Speed, T.P.: Comparison of discrimination methods for the classification of tumors using gene expression data. J. American Statistical Association 97(457), 77–87 (2002)
6. Dupuy, A., Simon, R.: Critical review of published microarray studies for cancer outcome and guidelines on statistical analysis and reporting. J. Natl. Cancer Institute 99(2), 147–157 (2007)
7. Freund, Y., Schapire, R.E.: Large margin classification using the perceptron algorithm. In: Proceedings of the 11th Annual Conference on Computational Learning Theory, pp. 209–217 (1998)
8. Friedman, H., Rafsky, L.C.: Multivariate generalization of the wald-wolfowitz and smirnov two-sample tests. Ann. Statist. 7, 697–717 (1979)

9. Giraud-Carrier, C., Vilalta, R., Brazdil, P.: Introduction to the special issue on meta-learning. Mach. Learn. 54(3), 187–193 (2004)
10. Golub, T.R., et al.: Molecular classification of cancer: class discovery and class prediction by gene expression monitoring. Science 286(5439), 531–537 (1999)
11. Hastie, T., Tibshirani, R., Friedman, J.: The elements of statistical learning: Data mining, inference and prediction. Springer, New York (2001)
12. Ho, T., Basu, M.: Complexity measures of supervised classification problems. IEEE Transactions on Pattern Analysis and Machine Intelligence 24(3), 289–300 (2002)
13. Irizarry, R.A., Warren, D., Spencer, F., Kim, I.F., Biswal, S., Frank, B.C., Gabrielson, E., Garcia, J.G.N., Geoghegan, J., Germino, G., Griffin, C., Hilmer, S.C., Hoffman, E., Jedlicka, A.E., Kawasaki, E., Martinez-Murillo, F., Morsberger, L., Lee, H., Petersen, D., Quackenbush, J., Scott, A., Wilson, M., Yang, Y., Ye, S.Q., Yu, W.: Multiple-laboratory comparison of microarray platforms. Nat. Methods 2(5), 345–350 (2005)
14. Kleinbaum, D.G., Klein, M.: Logistic Regression, 2nd edn. Springer, Heidelberg (2005)
15. Lorena, A.C., Costa, I.G., de Souto, M.C.P.: On the complexity of gene expression classification data sets. In: Proc. of the 8th International Conference on Hybrid Intelligent Systems, pp. 825–830. IEEE Computer Society Press, Los Alamitos (2008)
16. Lottaz, C., Kostka, D., Markowetz, F., Spang, R.: Computational diagnostics with gene expression profiles. Methods Mol. Biol. 453, 281–296 (2008)
17. McCallum, A., Nigam, K.: A comparison of event models for naive bayes text classification. In: AAAI/ICMC 1998 Workshop on Learning for Text Categorization, pp. 41–48 (1998)
18. Mitchell, T.: Machine Learning. McGraw-Hill, New York (1997)
19. Monti, S., et al.: Consensus clustering: a resampling-based method for class discovery and visualization of gene expression microarray data. Mach. Learn 52, 91–118 (2003)
20. Okun, O., Priisalu, H.: Dataset complexity in gene expression based cancer classification using ensembles of k-nearest neighbors. Artificial Intelligence in Medicine 45(2-3), 151–162 (2009)
21. Quackenbush, J.: Computational analysis of cDNA microarray data. Nature Reviews 6(2), 418–428 (2001)
22. Ramaswamy, S., et al.: Multiclass cancer diagnosis using tumor gene expression signatures. Proc. Natl. Acad. Sci. USA 98, 15149–15154 (2001)
23. Rosemblatt, F.: Principles of Neurodynamics: Perceptrons and the Theory of Brain Mechanisms. Spartan Books, New York (1962)
24. Slonim, D.: From patterns to pathways: gene expression data analysis comes of age. Nature Genetics 32, 502–508 (2002)
25. Smith, F.: Pattern classifier design by linear programming. IEEE Transactions on Computers 17(4), 367–372 (1968)
26. Sokal, R., Rohlf, F.: Biometry. W. H. Freeman and Company, New York (1995)
27. Spang, R.: Diagnostic signatures from microarrays: a bioinformatics concept for personalized medicine. BIOSILICO 1(2), 64–68 (2003)
28. Statnikov, A., Aliferis, C.F., Tsamardinos, I., Hardin, D., Levy, S.: A comprehensive evaluation of multicategory classification methods for microarray gene expression cancer diagnosis. Bioinformatics 21(5), 631–643 (2005)

29. van't Veer, L.J., Bernards, R.: Enabling personalized cancer medicine through analysis of gene-expression patterns. Nature 452(7187), 564–570 (2008)
30. Vapnik, V.N.: The nature of Statistical learning theory. Springer, New York (1995)
31. Witten, I.H., Frank, E.: Data Mining: Practical machine learning tools and techniques, 2nd edn. Morgan Kaufmann, San Francisco (2005)
32. Yeang, C.H., et al.: Molecular classification of multiple tumor types. In: Proc. 9th Int. Conf. on Intelligent Systems in Molecular Biology, vol. 1, pp. 316–322 (2001)
33. Zucknick, M., Richardson, S., Stronach, E.: Comparing the characteristics of gene expression profiles derived by univariate and multivariate classification methods. Statist. Appl. in Genetics and Molec. Biol. 7(1), 1–31 (2008)

Quantitative Improvements in cDNA Microarray Spot Segmentation

Mónica G. Larese[1,2,*] and Juan Carlos Gómez[1,2]

[1] Centro Internacional Franco-Argentino de Ciencias de la Información y de Sistemas CIFASIS-CONICET,
Bv. 27 de Febrero 210 Bis, 2000 Rosario, Argentina
[2] Laboratory for System Dynamics and Signal Processing, FCEIA, UNR,
Riobamba 245 Bis, 2000 Rosario, Argentina
larese@cifasis-conicet.gov.ar,
jcgomez@fceia.unr.edu.ar

Abstract. When developing a cDNA microarray experiment, the segmentation of individual spots is a crucial stage. Spot intensity measurements and gene expression ratios directly depend on the effectiveness and accuracy of the segmentation results. However, since the ground truth is unknown in microarray experiments, quantification of the accuracy of the segmentation process is a very difficult task. In this paper an improved unsupervised technique based on the combination of clustering algorithms and Markov Random Fields (MRF) is proposed to separate the foreground and background intensity signals used in the spot ratio computation. The segmentation algorithm uses one of two alternative methods to provide for initialization, namely K-means and Gaussian Mixture Models (GMM) clustering. This initial segmentation is then processed via MRF. Accuracy is measured by means of a set of microarray images containing spike spots where the target ratios are known *a priori*, thus making it possible to quantify the expression ratio errors. Results show improvements over state-of-the-art procedures.

Keywords: Gene expression, cDNA microarray segmentation, Gaussian Mixture Models, K-means clustering, Markov Random Field segmentation, quantitative segmentation errors.

1 Introduction

In spotted microarray experiments, segmentation tasks are necessary to separate the foreground and background intensity signals for individual spots, leading to the computation of the gene expression ratios. This is a fundamental task, since any inaccuracy arising from this step directly affects the subsequent analysis and conclusions about the expression profiles. The presence of noise and artifacts, defective spots and other sources of variability affect the quality of microarrays, thus making it necessary to use robust methods for segmentation.

* Author to whom all correspondence should be addressed.

K.S. Guimarães, A. Panchenko, T.M. Przytycka (Eds.): BSB 2009, LNBI 5676, pp. 60–72, 2009.

Many segmentation algorithms have been proposed in the last years, from the most traditional techniques used in the image analysis field to more advanced data mining procedures. However, their performance is usually evaluated without computing quantitative error measures (see e.g.: [1,2]). This is probably due to the fact that ground truth is completely unknown in real microarray images, *i.e.*, no *a priori* information is available about the true gene expression values.

In order to overcome subjective evaluation some authors [3,4,5] make use of simulated datasets to evaluate the performance of the algorithms in a pixel level basis. Simulated datasets are useful to evaluate pixel classification errors and to detect spatial structure errors introduced by the algorithms. Simulated spots are constructed by assigning a target ratio to each one of them. In this case a model of the microarray is assumed in order to generate the simulation, *i.e.*, spot shapes, intensity distributions, noise models, etc.

The segmentation of real microarrays is much more difficult, since real physical phenomena are involved. In this situation control or spike spots with *a priori* fixed target ratios are needed in the microarray construction. Those target ratios are used to compare different segmentation methods, as in [6].

In this paper, a completely unsupervised segmentation algorithm is proposed and evaluated in terms of quantitative error measures on real microarray images. Relative improvements over state-of-the-art methods are computed. The method consists of an initial segmentation step performed on the spot intensities followed by an unsupervised Markov Random Field (MRF) segmentation procedure. A two-threshold percentile method is then used to filter outlier intensities, leading to compute the expression ratios.

Two alternative approaches are considered to perform the initialization step: K-means clustering and Gaussian Mixture Models (GMM). The performances of both initialization techniques are quantitatively compared. Results show important improvements over state-of-the-art algorithms.

The rest of the paper is organized as follows. The real dataset used for the analysis is briefly described in Section 2. The microarray segmentation procedure is detailed in Section 3. The two initialization schemes, namely the K-means and GMM unsupervised segmentation algorithms, are presented in Subsections 3.1 and 3.2, respectively. The MRF segmentation technique is explained in Subsection 3.3, whereas the expression ratio calculation and error measurements are commented in Subsections 3.4 and 3.5, respectively. The results are presented and discussed in Section 4. Finally, some conclusions are drawn in Section 5.

2 Dataset Description

The dataset used in this paper consists of sixteen real two-channel microarray images designed for cancer research purposes at the Genomic Medicine Research Core Laboratory of Chang Gung Memorial Hospital, in Taiwan. They are grouped into eight pairs of dye swapped microarrays, denoted as (1-1s) to (8-8s).

Each microarray is composed by 32 subgrids, each one containing 22 rows by 22 columns of spots. The lower 16 subgrids in each microarray contain replicated spots from the upper 16 subgrids. The reader is referred to [6,7,8] for more details about the microarray experiment and studies carried out on them.

There is a total of 4096 spike spots in the whole experiment, for which true intensity ratios are known. Each microarray contains 256 spike spots, 8 spikes per subgrid. They are located at column 22, from rows 3 to 10. For microarrays (1-1s) to (4-4s), the Cy5/Cy3 ratios for the 8 spikes in each subgrid are [0.1, 0.1, 0.2, 0.2, 0.4, 0.4, 1.0, 1.0]. For microarrays (5-5s) to (8-8s), all the spikes have a constant ratio of 0.2.

3 Microarray Segmentation Procedure

In this work, spike spots are individually segmented using a local two-step procedure. The first step is the initialization and it consists of a preliminary partitioning of the spot intensity distribution into two groups, thus clustering the pixels into foreground and background. This initial segmentation result is then used to initialize the MRF model in order to accelerate convergence.

The initialization step assumes that the intensities of the pixels are independently distributed, and that there is no spatial relations between them. The MRF incorporates the spatial information taking into account the vicinity of the pixels and modeling that relation under a Bayesian framework.

Manual or automatic procedures can be used to initialize the MRF. For example in [3], the authors state that trying several automatic thresholding techniques did not give satisfactory results. Thus, they employed the percentile method to do initial labeling. In this paper, two alternative automatic schemes are proposed to perform the MRF initial segmentation: the K-means and the GMM clustering techniques.

K-means and GMM have previously been used in the context of microarray segmentation. In Bozinov *et al.* [9], the information from the two channels is used to construct the feature vectors which are then clustered using the K-means algorithm. In [1], K-means and fuzzy K-means are used as single methods to cluster the spot intensities. Regarding GMM, it was proposed in Blekas *et al.* [10] to develop spot segmentation. Later on, Chen *et al.* [6] applied both single GMM and GMM in combination with Kernel Density Estimation (KDE) in order to improve segmentation and therefore gene expression estimation.

However, to the best of the present authors' knowledge, K-means and GMM have not been used yet as an initialization stage for MRF for microarray segmentation tasks. K-means and GMM techniques are briefly explained in the following two Subsections. MRF based segmentation is detailed in Subsection 3.3.

3.1 K-means Initial Segmentation

K-means is a fast and simple very well-known partitional clustering algorithm [11]. Given a dataset $\mathbf{D} = \{x_1, x_2, ..., x_N\}$, a number K of clusters and a metric (usually the Euclidean distance), the algorithm initializes a set of centers $M = \{\mu_1, \mu_2, ..., \mu_K\}$ at random. Then, it divides $\mathbf{D}$ into K disjoint subsets and assigns each x_i to a cluster according to the nearest center. The center locations are updated iteratively by minimizing the distortion in all the clusters. The process is iterated until there are no further changes in M using the Expectation-Maximization (EM) algorithm.

In the context of this paper, K-means is used to cluster the 1-D spot pixel intensities dataset $\mathbf{D}$ into two groups ($K = 2$): foreground (highest intensities) and background (lowest intensities). As the spike spots in the dataset considered in this work are known to have the foreground mean higher than the background mean, this hypothesis is used in this paper, which is also commonly used in the literature [6,12]. In future work the procedure should be extended to consider spots where the background signal is higher than the foreground.

The two microarray channels are processed separately. Some target spots present very few (only one or two) extremely high intensities, giving rise to a non-representative foreground cluster. For this special case, K is set to 3, and the clusters corresponding to the highest and middle intensities are merged into foreground, while the cluster with the lowest intensities is assigned to background.

3.2 Gaussian Mixture Model (GMM) Initial Segmentation

GMM is a classic clustering algorithm that allows to model a multivariate distribution of independently drawn patterns as the sum (mixture) of a finite number $k = 2, ..., K$ of multivariate Gaussian distributions [13].

In this paper, GMM is used to model the 1-D dataset $\mathbf{D} = \{x_1, x_2, ..., x_N\}$ of the N pixel intensities for an individual spot. Therefore, the spot intensity distribution is modeled as the mixture Y of the foreground and background Gaussian distributions ($K = 2$) that best fit the observed intensities, using as a convention that the foreground mean is higher than the background mean, as discussed above.

Let $\phi_{\theta_i}(x)$ be the normal density with parameters $\theta_i \triangleq (\mu_i, \sigma_i^2)$, $i = 1, 2$. The density of Y is then as in Eq. (1), where π is the mixing probability parameter.

$$p_Y(x) = (1 - \pi)\phi_{\theta_1}(x) + \pi\phi_{\theta_2}(x) \tag{1}$$

In order to fit the model to the observed data, maximum likelihood is used. The optimization is performed resorting to the EM algorithm. The parameters to be optimized are

$$\theta \triangleq (\pi, \theta_1, \theta_2) = (\pi, \mu_1, \sigma_1^2, \mu_2, \sigma_2^2) \tag{2}$$

In order to take into account the presence of spots with extremely bright pixels, leading to non-representative foreground clusters, a procedure similar to the one employed in the K-means initialization approach (Subsection 3.1) is also applied here. The intensity distribution is modeled by extending the above formulae to the mixture of three Gaussians instead of two, and the pixels corresponding to the two Gaussian distributions with the highest means are merged into the foreground cluster.

3.3 Markov Random Field (MRF) Segmentation

The result of the initialization step is a binary image where background and foreground pixels are labeled with 0s and 1s, respectively. In this initial segmentation the intensities of the pixels are assumed independently distributed and no spatial relations among them are considered.

The purpose of the MRF stage is to improve the initial segmentation and take into account the local interaction of pixels. An MRF model [14] is designed to estimate the intensity distribution over the $M \times N$ lattice of pixels, in lexicographic order denoted by $\mathcal{S} = \{s_1, ..., s_{M\times N}\}$, using an 8-pixel neighborhood. The set of possible labels for each site (pixel) is $\mathcal{L} = \{0, 1\}$.

Let the random variable X be the process of assigning a value from $\mathcal{L}$ to site s_i. Then, x_{s_i} denotes the label at pixel s_i. The Hammersly-Clifford theorem states that the density of X is given by the Gibbs density as

$$\begin{aligned} p(x) &= Z^{-1} e^{-\beta U(x)}, \\ U(x) &= \sum_{c \in \mathcal{C}} V_c(x_c) \end{aligned} \tag{3}$$

In the formula above, the set $\mathcal{C}$ is the set of *cliques* for every site. A clique is the set of pixels that are all neighbors of each other under the established neighborhood system. The positive parameter β controls the size of the clusters, Z is a normalization constant (the so-called partition function) and U is the energy function. The V_c are the two-point potentials, defined as

$$V(x) = \begin{cases} -1 & x_i = x_j, \text{ with } i, j \in \mathcal{C} \\ 1 & x_i \neq x_j, \text{ with } i, j \in \mathcal{C} \end{cases} \tag{4}$$

Let f be the observed spot image, which is a realization of the random field F. Let x^{TRUE} be the true (unknown) labeling of pixels of f, and $\hat{x}$ its estimate. Under a Bayesian framework, Eq. (5) describes the posterior probability

$$p(x|f) \propto p(f|x)p(x) \tag{5}$$

where $p(f|x)$ is the likelihood and $p(x)$ is the prior distribution for x^{TRUE}. In this paper, the likelihood given class x_s (background or foreground) is modeled as Gaussian, as shown below

$$p(f|x_s) = \frac{1}{\sqrt{2\pi\sigma_{x_s}^2}} e^{-\frac{(f-\mu_{x_s})^2}{2\sigma_{x_s}^2}} \tag{6}$$

where μ_{x_s} and σ_{x_s} are the mean and standard deviation, respectively, for pixels labeled with x_s (0 and 1 for background and foreground, respectively). The estimated labeling $\hat{x}$ was computed through Maximum a Posteriori (MAP) estimation by means of minimizing the negative of the log of the posterior via Iterated Conditional Modes (ICM) [15], as depicted in Eq. (7). The values of μ_{x_s} and σ_{x_s} are computed at each iteration according to the current labeling.

$$\hat{x} = \arg\max_{x} \{\log p(f|x) + \log p(x)\} =$$
$$= \arg\min_{x_s \in \mathcal{L}} \left[\frac{1}{2}\log(2\pi\sigma_{x_s}^2) + \frac{(f-\mu_{x_s})^2}{2\sigma_{x_s}^2} + \beta U(x_s) \right] \quad (7)$$

In this paper, β was set to 0.01. Four to seven ICMs were considered and the corresponding segmentation results were compared, as it is described in Section 4. The initial values for x_s, μ_{x_s} and σ_{x_s} were obtained from the initial labeling resulting from the K-means and GMM segmentation algorithms explained in Subsections 3.1 and 3.2, respectively.

3.4 Expression Ratios Calculation

The segmented spike spots obtained from the MRF step are used to compute the background subtracted expression ratios. In order to increase the robustness of the method against outliers, intensity values from the foreground and background between $[p_1, p_2]$ percentiles of each class distribution are used to compute the Cy5/Cy3 ratios for each spike spot.

The percentile thresholding method [16] is a well-known thresholding technique in the image processing field. In this paper, a two-threshold technique with percentile values p_1 and p_2 was selected. Several sets of percentiles were tested, obtaining the best results (in terms of the quantitative error measurements) for the sets $[p_1 = 10\text{th}, p_2 = 90\text{th}]$ and $[p_1 = 15\text{th}, p_2 = 85\text{th}]$. These results are analyzed in Section 4.

3.5 Quantitative Accuracy Assessment

The accuracy of the segmentation process is analyzed for the K-means and GMM initialization approaches, and for the two different percentile ranges used for ratio computation. Two error measurements, namely the Sum of Squared Errors (SSE) and the Sum of Squared Relative Errors ($SSRE$), also used by Chen *et al.* in [6], are computed for each one of the 16 microarrays in the dataset. These two errors are also used to compare the method proposed in this paper to the GKDE (GMM in combination with KDE), KDE and GMM approaches in [6] and to the commercial software GenePix 6.0[1]. Among the several adaptive segmentation methods available in Genepix, the irregular one is the most accurate according to the study provided by [6]. That is the reason why it was chosen for comparison in this paper. The SSE and $SSRE$ are described in Eqs. (8) and (9), respectively, where $P = 8$ stands for the number of spikes per subgrid, $Q = 32$ for the number of subgrids in a single microarray, and $\hat{r}$ and r are the estimated and true expression ratios, respectively. True expression ratios correspond to the known ratios for spike spots, as described in Section 2.

$$SSE = \sum_{p=1}^{P}\sum_{q=1}^{Q}(\hat{r}_{p,q} - r_{p,q})^2 \quad (8)$$

[1] http://www.moleculardevices.com/pages/software/gn_genepix_pro.html

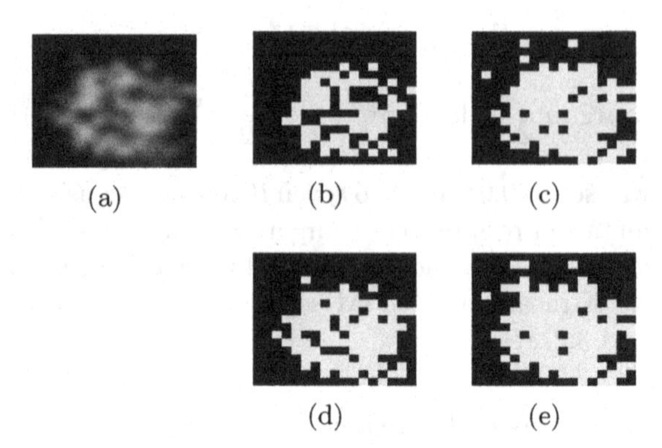

Fig. 1. Segmentation examples for an individual spot from the Cy5 channel. (a) Original spot. (b) K-means segmentation. (c) MRF after K-means segmentation. (d) GMM segmentation. (e) MRF after GMM segmentation.

$$SSRE = \sum_{p=1}^{P} \sum_{q=1}^{Q} \left[\frac{(\hat{r}_{p,q} - r_{p,q})}{r_{p,q}} \right]^2 \tag{9}$$

The percentage of relative improvement RI of the proposed method over GKDE, KDE, GMM and Genepix 6.0 Irregular, is computed for each microarray as

$$RI = \frac{(Error_{\text{other}} - Error_{\text{proposed}})}{Error_{\text{other}}} * 100 \tag{10}$$

where $Error_{\text{other}}$ stands for the SSE or $SSRE$ of any of the state-of-the-art methods, and $Error_{\text{proposed}}$ is the SSE or $SSRE$ of the proposed algorithm.

4 Discussion

The segmentation result for an individual spike spot from the Cy5 channel is shown in Fig. 1. Figures 1(b) and 1(c) display the segmented spot after initialization via K-means clustering and application of MRF segmentation, respectively. Figures 1(d) and 1(e) depict the result of using GMM for initialization and then applying the MRF method. As it can be seen from this example, the MRF segmentation step helps to improve the detection of foreground and background, and different initialization schemes produce different final results from the MRF.

Figures 2 and 3 show the medians (along the 16 microarrays) of the relative improvements (RIs) measured by the SSEs and the $SSRE$s, respectively, introduced by the proposed method over four state-of-the-art procedures, namely GKDE, KDE, GMM and Genepix Irregular. Figures 2(a) and 3(a) reflect the improvements using the K-means initialization approach, while Figs. 2(b) and 3(b) correspond to the GMM initialization technique before applying the MRF segmentation step. The vertical dashed blue line in each graphic separates the results

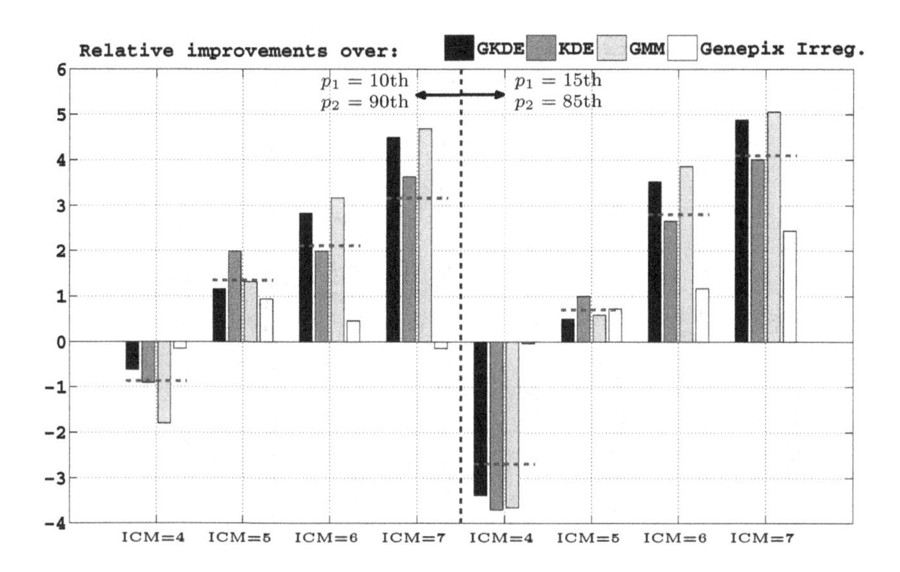

(a)

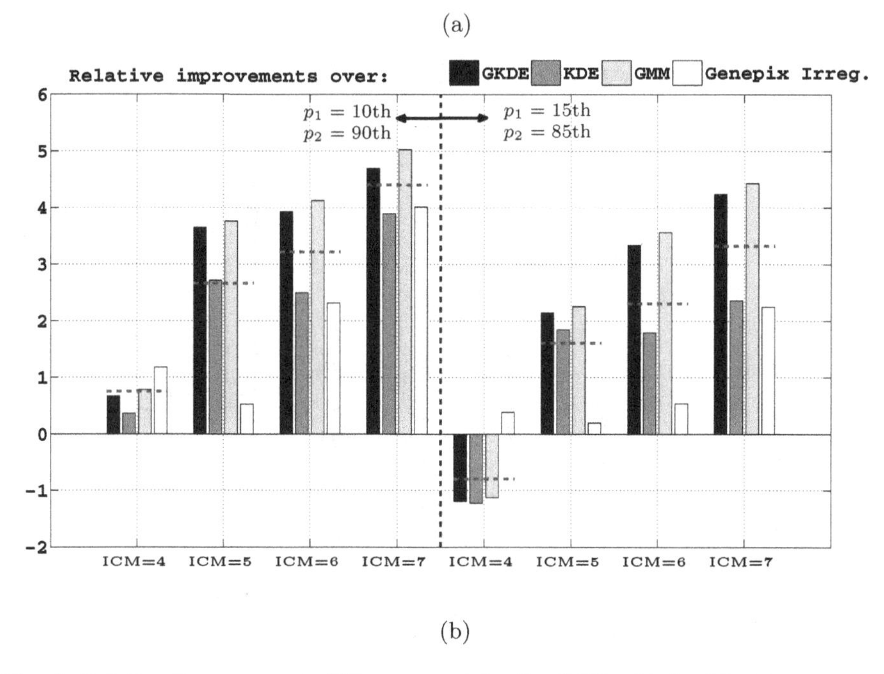

(b)

Fig. 2. Relative improvements (%) measured by the *SSE*s obtained by the proposed method over GKDE, KDE and GMM [6] and the commercial software GenePix 6.0 (irregular configuration): (a) Using the K-means initialization approach before the MRF segmentation. (b) Using the GMM initialization approach before the MRF segmentation. (See the text for a detailed explanation).

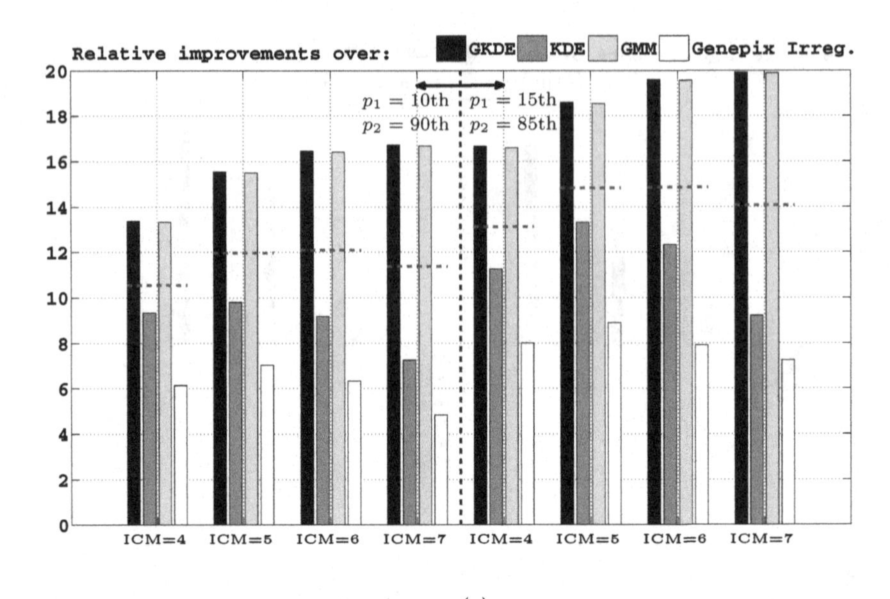

(a)

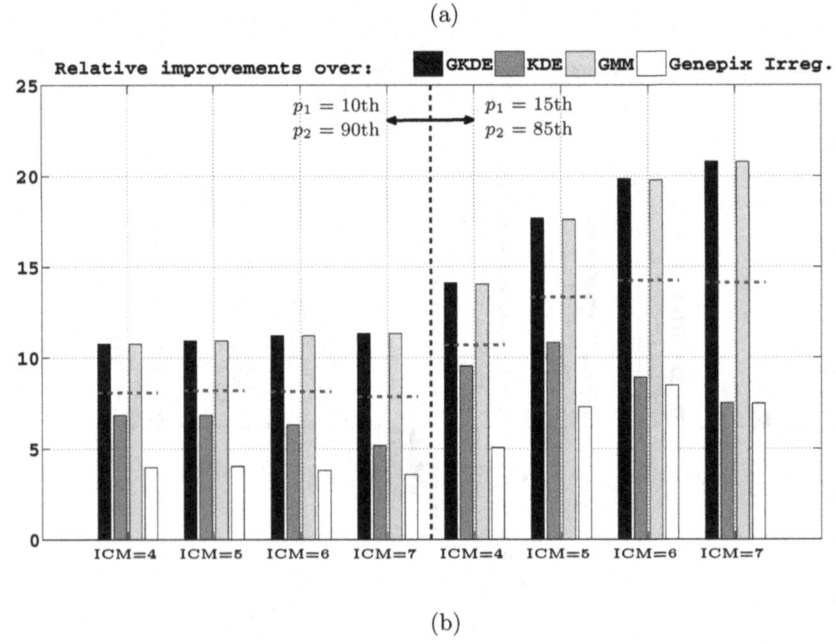

(b)

Fig. 3. Relative improvements (%) measured by the *SSRE*s obtained by the proposed method over GKDE, KDE and GMM [6] and the commercial software GenePix 6.0 (irregular configuration): (a) Using the K-means initialization approach before the MRF segmentation. (b) Using the GMM initialization approach before the MRF segmentation. (See the text for a detailed explanation).

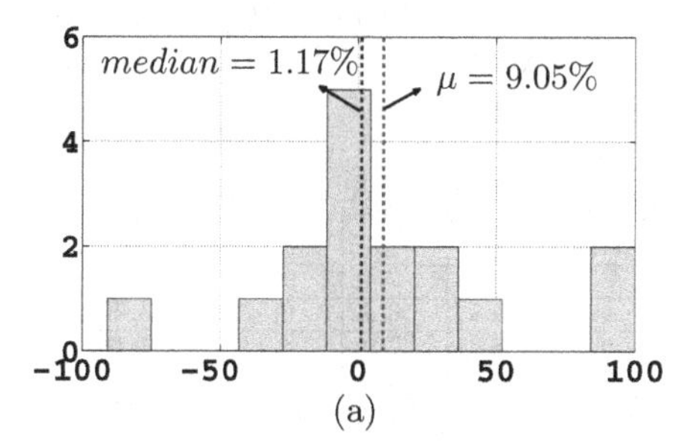

(a)

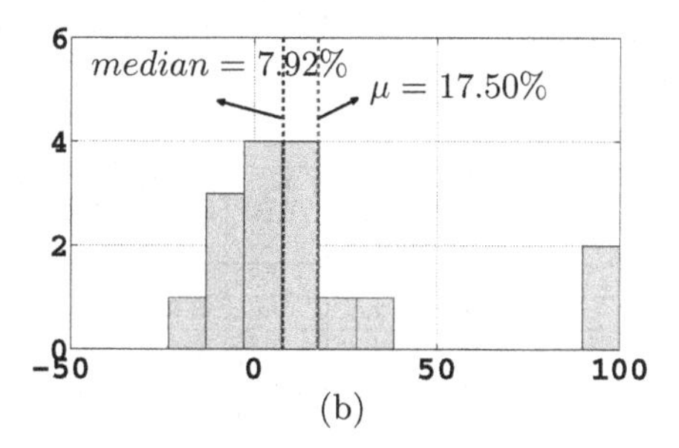

(b)

Fig. 4. Relative improvements (%) frequency distribution of the proposed method over Genepix Irregular measured by (a) SSEs, and (b) $SSRE$s

for the two different percentile ranges used to threshold the intensities in the expression ratio calculation, *i.e.*, $[p_1 = 10\text{th}, p_2 = 90\text{th}]$ and $[p_1 = 15\text{th}, p_2 = 85\text{th}]$. The horizontal dashed red lines represent the mean RIs in each group of bars.

The results from Figs. 2 and 3 show that taking more than 4 ICMs leads to positive RIs in all the considered cases. The proposed method achieves the highest improvements over GKDE and GMM state-of-the-art methods. Up to a 5% of improvement over GMM (measured by the SSE) is obtained using 7 ICMs, the K-means initialization approach, $p_1 = 15$th and $p_2 = 85$th. An RI of 20.8% over GKDE and GMM (measured by the $SSRE$) is achieved using 7 ICMs, the GMM initial labeling, $p_1 = 15$th and $p_2 = 85$th. The smallest improvement is obtained over Genepix Irregular.

The mean values of the RIs for the 4 methods (measured by the SSEs) tend to increase according to the ICM values. However, the mean RIs for the 4 methods (measured by the $SSRE$s) seem to grow up to ICM=6, and then they start decreasing. The RIs are also slightly higher for $p_1 = 15$th, $p_2 = 85$th and the K-means initialization approach, except for the data in Fig.2(b).

Tables 1 and 2 show the SSEs and $SSRE$s, respectively, in addition to the corresponding RIs for each microarray image, as well as the median RIs along the 16 microarrays. The proposed method used for comparison is a K-means initialization approach, followed by MRF optimized by means of 6 ICMs, and $[p_1 = 15\text{th}, p_2 = 85\text{th}]$. As it can be observed from Tables 1 and 2, some outlier SSEs and $SSRE$s arise, *e. g.*, SSE and $SSRE$ for Genepix and microarray 7s. These huge numbers arise when Genepix estimates many true constant 0.2 ratios as negative ratios with absolute values greater than 1.

The median is a robust estimator of central tendency, and it is preferable over the mean in the presence of small sample sizes, skewed distributions, existence of outliers and existence of extreme values. It is a more representative estimator of where the middle of the distribution lies. The median along the 16 microarrays for each one of the two RI distributions (measured by the SSE and $SSRE$) from the example for Genepix Irregular (last columns of Tables 1 and 2) can be observed in Figures 4(a) and 4(b). The mean, instead, is not a representative estimator.

From Tables 1 and 2 it can also be observed that, as it is expected from Eq. (8) and Eq. (9), the RIs measured by the SSEs and $SSRE$s are the same

Table 1. Sum of Square Errors (SSEs) and Relative Improvements (RIs) of the proposed algorithm over state-of-the-art methods, for ICM=6, p_1 = 15th, p_2 = 85th and the K-means initialization approach before MRF segmentation

Microarray	Sum of Square Errors (SSEs) : Proposed method	GKDE	KDE	GMM	Genepix Irreg.	Relative improvements over: GKDE (%)	KDE (%)	GMM (%)	Genepix Irreg. (%)
1	2.85	2.87	2.78	2.87	3.02	0.62	-2.48	0.66	5.73
2	2.83	3.02	3.02	3.03	4.28	6.47	6.32	6.57	33.90
3	4.98	5.43	5.41	5.44	5.04	8.31	7.91	8.43	1.18
4	20.05	9.39	9.29	9.70	17.44	-113.47	-115.80	-106.67	-14.92
5	0.28	0.41	0.32	0.42	0.28	33.05	12.71	33.70	1.16
6	0.29	0.31	0.31	0.31	0.29	4.35	5.59	4.66	-2.27
7	1.77	2.44	2.38	2.44	174.11	27.21	25.34	27.24	98.98
8	3.53	4.44	4.08	4.44	3.77	20.51	13.43	20.53	6.52
1s	4.99	4.41	3.46	4.40	2.62	-13.12	-44.14	-13.53	-90.91
2s	7.00	2.06	2.68	2.27	5.19	-239.45	-161.66	-209.02	-34.81
3s	11.65	12.31	14.82	12.31	19.30	5.37	21.39	5.38	39.66
4s	79.45	88.79	99.96	86.53	119.15	10.52	20.52	8.19	33.32
5s	0.62	0.49	0.48	0.49	0.58	-26.22	-27.26	-25.96	-6.13
6s	0.26	0.27	0.26	0.27	0.25	2.70	-0.27	3.06	-3.28
7s	0.64	0.51	0.50	0.51	8695318290.27	-26.40	-29.46	-26.16	100.00
8s	0.48	0.40	0.40	0.40	0.39	-20.03	-19.43	-19.73	-23.35
Median						3.53	2.66	3.86	1.17

Table 2. Sum of Square Relative Errors ($SSRE$s) and Relative Improvements (RIs) of the proposed algorithm over state-of-the-art methods, for ICM=6, p_1 = 15th, p_2 = 85th and the K-means initialization approach before MRF segmentation

Microarray	Sum of Square Relative Errors ($SSRE$s) : Proposed method	GKDE	KDE	GMM	Genepix Irreg.	Relative improvements over: GKDE (%)	KDE (%)	GMM (%)	Genepix Irreg. (%)
1	52.24	85.48	82.50	85.48	76.96	38.89	36.67	38.89	32.12
2	40.45	55.01	45.90	55.03	43.37	26.47	11.87	26.49	6.73
3	53.99	80.42	77.15	80.42	71.23	32.87	30.02	32.87	24.20
4	53.87	29.86	28.02	30.17	51.54	-80.40	-92.24	-78.55	-4.51
5	6.90	10.40	7.91	10.41	6.98	33.70	12.80	33.76	1.16
6	7.29	7.61	7.73	7.65	7.13	4.10	5.64	4.62	-2.27
7	44.33	60.91	59.38	60.92	4352.74	27.22	25.35	27.23	98.98
8	88.21	110.99	101.91	110.99	94.36	20.52	13.44	20.53	6.52
1s	25.60	33.01	31.74	32.98	29.75	22.42	19.33	22.36	13.95
2s	24.66	26.90	27.21	26.91	27.12	8.34	9.39	8.36	9.10
3s	106.66	149.07	130.20	149.74	123.46	28.45	18.08	28.77	13.61
4s	548.85	675.01	648.21	674.39	631.79	18.69	15.33	18.62	13.13
5s	15.40	12.20	12.11	12.22	14.51	-26.20	-27.20	-25.99	-6.13
6s	6.57	6.78	6.55	6.79	6.36	3.15	-0.27	3.22	-3.28
7s	16.09	12.71	12.43	12.74	217382957256.80	-26.60	-29.46	-26.27	100.00
8s	11.97	9.91	10.03	10.00	9.71	-20.82	-19.43	-19.76	-23.35
Median						19.61	12.34	19.57	7.92

for microarrays (5,5s) to (8,8s), where the true intensity ratios are constantly equal to 0.2.

5 Concluding Remarks

Quantitative evaluation of the segmentation step is fundamental to compare and assess the accuracy of the different algorithms for microarray segmentation.

In this paper a new segmentation algorithm is proposed, based on MRF segmentation previously initialized via two alternative approaches, namely K-means and GMM clustering. The algorithm is compared to state-of-the-art methods in terms of quantitative measures (SSE and $SSRE$) that allow to compute the relative improvements on background subtracted ratio estimates. A database consisting on real microarray images with *a priori* known spike spots ratios is used to quantitatively assess the accuracy.

The results demonstrate that the proposed algorithm obtains very good results in all the tested configurations for more than four ICMs in the MRF optimization step. The highest improvements are obtained over GKDE and GMM methods, while the smallest improvements are over Genepix. However, for a particular single image the computed errors for Genepix are extremely high, whereas the errors obtained by the proposed algorithm are much more reasonable. Current work is being developed in order to test the performance of the proposed algorithms on artificial microarray images generated by the model in [5].

References

1. Wang, Y.P., Gunampally, M.R., Cai, W.W.: Automated segmentation of microarray spots using fuzzy clustering approaches. In: IEEE Workshop on Machine Learning for Signal Processing, pp. 387–391 (2005)
2. Gottardo, R., Besag, J., Stephens, M., Murua, A.: Probabilistic segmentation and intensity estimation for microarray images. Biostatistics 7(1), 85–99 (2006)
3. Demirkaya, O., et al.: Segmentation of cDNA microarray spots using Markov Random Field modeling. Bioinformatics 21(13), 2994–3000 (2005)
4. Lehmussola, A., Ruusuvuori, P., Yli Harja, O.: Evaluating the performance of microarray segmentation algorithms. Bioinformatics 22(23), 2910–2917 (2006)
5. Nykter, M., Aho, T., et al.: Simulation of microarray data with realistic characteristics. BMC Bioinformatics 7(349), 1–17 (2006)
6. Chen, T.B., Lu, H.H.S., et al.: Segmentation of cDNA microarray images by kernel density estimation. J. of Biomedical Informatics 41, 1021–1027 (2008)
7. Wang, T., Lee, Y., et al.: Establishment of cDNA microarray analysis at the Genomic Medicine Research Core Laboratory (GMRCL) of Chang Gung Memorial Hospital. Chang Gung Med. Journal 27(4), 243–260 (2004)
8. Chao, A., Wang, T.H., et al.: Molecular characterization of adenocarcinoma and squamous carcinoma of the uterine cervix using microarray analysis of gene expression. Int. J. Cancer 119(1), 91–98 (2006)
9. Bozinov, D., Rahnenfürher, J.: Unsupervised technique for robust target separation and analysis of DNA microarray spots through adaptive pixel clustering. Bioinformatics 18(5), 747–756 (2002)
10. Blekas, K., Galatsanos, N.P., et al.: Mixture model analysis of DNA microarray images. IEEE Transactions on Medical Imaging 24(7), 901–909 (2005)
11. Bishop, C.: Pattern Recognition and Machine Learning. Springer, Heidelberg (2006)

12. Blekas, K., Galatsanos, N.P., et al.: An unsupervised artifact correction approach for the analysis of DNA microarray images. In: IEEE ICIP, Barcelona, pp. 165–168 (2003)
13. Friedman, J., Hastie, T., Tibshirani, R.: The Elements of Statistical Learning: Data Mining, Inference and Prediction. Springer, Heidelberg (2001)
14. Geman, S., Geman, D.: Stochastic relaxation, Gibbs distributions and the Bayesian restoration of images. IEEE Trans. on P.A.M.I. 6, 721–741 (1984)
15. Besag, J.: On the statistical analysis of dirty images. J. of the Royal Statistical Soc. Series B 48(3), 259–302 (1986)
16. Sonka, M., Hlavac, V., Boyle, R.: Image processing analysis and machine vision. Thomson (2008)

SOM-PORTRAIT: Identifying Non-coding RNAs Using Self-Organizing Maps

Tulio C. Silva[1], Pedro A. Berger[1], Roberto T. Arrial[2], Roberto C. Togawa[3], Marcelo M. Brigido[2], and Maria Emilia M.T. Walter[1]

[1] Department of Computer Science - Institute of Exact Sciences
[2] Laboratory of Molecular Biology - Institute of Biology
Campus Universitario Darcy Ribeiro, University of Brasilia, Zip Code 70910-900
[3] Bioinformatics Laboratory, EMBRAPA Genetic Resources and Biotechnology, Zip Code 70770-900
Brasilia-Brazil

Abstract. Recent experiments have shown that some types of RNA may control gene expression and phenotype by themselves, besides their traditional role of allowing the protein synthesis. Roughly speaking, RNAs can be divided into two classes: mRNAs, that are translated into proteins, and non-coding RNAs (ncRNAs), which play several cellular important roles besides protein coding. In recent years, many computational methods based on different theories and models have been proposed to distinguish mRNAs from ncRNAs. Particularly, Self-Organizing Maps (SOM), a neural network model, is time efficient for the training step, and present a straightforward implementation that allow easily increasing of the number of classes for clustering the input data. In this work, we propose a method for identifying non-coding RNAs using Self Organizing Maps, named SOM-PORTRAIT. We implemented the method and applied it to a data set containing Assembled ESTs of the *Paracoccidioides brasiliensis* fungus transcriptome. The obtained results were promising, with the advantage that the time and memory requirements needed to our SOM-PORTRAIT are much less than those needed for methods based on the Support Vector Machine (SVM) paradigm, like PORTRAIT.

1 Introduction

The usual view of the central dogma of molecular biology [25] predicts that genetic information flow from DNA to proteins using RNA as intermediate. DNA is responsible for the genotype of a cell while protein is responsible for the cell's phenotype. This orthodox view of the central dogma suggests that RNA is an auxiliary molecule involved in all stages of protein synthesis and gene expression. But recent experiments have shown that some types of RNA may indeed control gene expressing and phenotype by themselves. Many other biological functions of RNAs are already known, and new functions are continuously being discovered. Roughly speaking, RNAs can be divided into two classes, mRNAs - which are

K.S. Guimarães, A. Panchenko, T.M. Przytycka (Eds.): BSB 2009, LNBI 5676, pp. 73–85, 2009.

translated into proteins, and non-coding RNAs (ncRNAs) - which play several cellular important roles besides protein coding.

In recent years, many computational methods have been proposed to distinguish mRNAs from ncRNAs. It is noteworthy that traditional methods that successfully identify mRNAs in general fail when used to identify ncRNAs, although they can be somewhat combined with other methods to identify a few number of well conserved RNAs, like rRNA. For example, BLAST [1] with customized parameters together with covariance models approaches correctly identified snRNAs [20]. Then, methods based on different theories have been developed, such as theory of probability like Infernal [6], thermodynamics [26,11], or Support Vector Machine (SVM) like CONC [17], CPC [14] and PORTRAIT [4].

Particularly, although CONC [17] and PORTRAIT [4] had achieved good results, a potential drawback of an algorithm based on the SVM model when dealing with large number of RNA sequences is both training time and memory requirement. For example, when n training instances must be held in memory, the best-known SVM implementation takes $O(n^a)$ time, with a typically between 1.8 and 2. But there is another neural network model, the Self-Organizing Maps (SOM), that takes $O(n)$ time for the training step, and has a straightforward implementation which allow easily increasing of the number of classes for grouping the input data, while this is far more complicated for the SVM approach.

In this context, the objective of this work is to propose a method for identifying ncRNAs using Self Organizing Maps, named SOM-PORTRAIT. In Section 2, we briefly discuss ncRNAS and our previous program PORTRAIT, that is based on the SVM model. In Section 3 we shortly describe the SOM model. In Section 4 we present our SOM model to identify ncRNAs, and show some implementation details. In Section 5, we present the results of applying our method to the *Paracoccidioides brasiliensis* fungus and compare them with SVM-PORTRAIT. Finally, in Section 6, we conclude and suggest future work.

2 About ncRNAs and SVM-PORTRAIT

As said before, experimental evidences have been suggesting that most of RNA transcribed throughout the genome does not code for proteins [19]. These RNAs have been called ncRNAs, a heterogeneous category of RNAs that includes regulatory molecules, conserved molecules with unknown function and transcriptional noise [24]. So, ncRNAs is an expanding class that includes different types of RNA involved in several cell activities, such as the well known ribosomal RNA (rRNA) and the transfer RNA (tRNA), both involved in the protein biosynthetic machinery, but it also includes small nuclear RNA (snRNA), small nucleolar RNA (snoRNA) and micro RNA (miRNA), among others. A special group is composed by a mRNA like ncRNA, that is, these molecules act very alike mRNA, but does not code for proteins, and frequently contaminate mRNA preparation [8].

By definition, ncRNA is characterized by the absence of an open reading frame (ORF), but this premise is misleading, since the fortuitous presence of small ORFs is quite frequent. On the other side, the absence of a detectable

ORF may not be observed in low quality transcripts, as those normally found in expressed sequence tag (ESTs). Therefore, to detect ncRNA from a set of ESTs composing a transcriptome is a challenging task.

In order to distinguish coding and non-coding RNA from an EST data set, we had previously proposed the PORTRAIT program [4], based on the SVM model, from here after named SVM-PORTRAIT. This program focused on small transcriptome projects, based in EST derived from poorly characterized organisms, for which there is just a little genetic information. SVM-PORTRAIT relies on intrinsic features of RNA sequence and represents an improvement for the CONC [18] and CPC [14] programs, since it does not use homology derived information associated to an error prone translator for ORF definition. SVM-PORTRAIT uses the LIBSVM [5] v2.84 implementation, with Radial Basis Function kernel, set as C-SVM and binary classification problem, and creates two classes, coding (positive set) and non-coding (negative set) RNA. Two models were induced separately: a protein-dependent one, induced with dbTR_OP set as training data, and a nucleotide-only using dbTR_OA for training.

3 Self-Organizing Maps

A Self-Organizing Map (SOM) is an Artificial Neural Network first described by T. Kohonen [12], also called Kohonen Map or Kohonen Neural Network. A SOM represents an open set of multivariate items by a finite set of model items, which makes it useful for classifying high-dimensional data [22]. A SOM has a simple organization (Figure 1) composed by only two layers: the input layer (not computational) and the output layer, also known as the Kohonen layer.

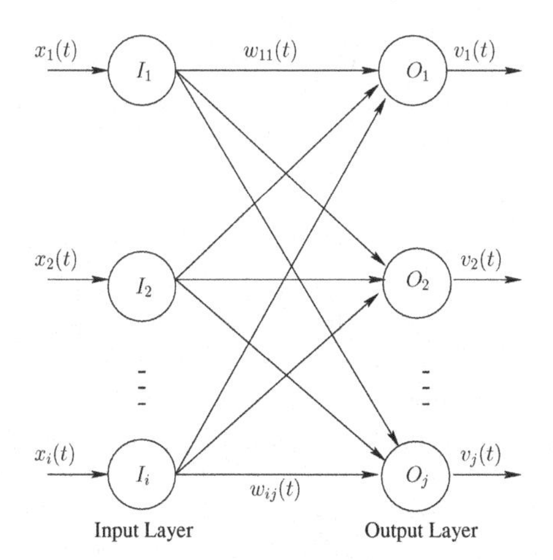

Fig. 1. The layers and vectors of a SOM: input vector $\bar{x}(t) = [x_1(t)x_2(t)\ldots x_i(t)]$, weight vector $\bar{w}_j(t) = [w_{1,1}w_{1,2}\ldots w_{i,j}]$ and activation level $v_j(t)$

When a stimulus $x_i(t)$ is presented, neurons "compete" by mutual lateral inhibition. The winner neuron has the higher activation level $v_j(t)$. So, the activation level $v_j(t)$ can be expressed as follows, in which $w_{i,j}$ is the ith element of the weight vector $\bar{w}_j(t)$, $v_j(t)$ is the activation level of the jth neuron, and $x_i(t)$ is the ith element of the input vector $\bar{x}(t)$:

$$v_j(t) = w_{1,1}(t)x_1(t) + w_{2,1}(t)x_2(t) + \ldots + w_{i,j}(t)x_i(t)$$

The training formula for a neuron with weight vector $\bar{w}_j(t)$ is given as follows, in which $\alpha(t)$ is the learning coefficient that is gradually reduced and $h_j(t)$ is the neighborhood function:

$$\bar{w}_j(t+1) = \bar{w}_j(t) + h_j(t)\alpha(t)\left[\bar{x}(t) - \bar{w}_j(t)\right]$$

During the training, the neighborhood function $h_j(t)$ depends on the distance between the winner neuron and the jth neuron. This function could be as simple as a constant for a number of neurons close enough to the winner neuron, but an widely used choice is a gaussian function [10]. The neighborhood function shrinks from one training epoch to another.

The training process is repeated for all input vectors for a number of interactions. The network associate output nodes with clusters of input data sharing some characteristics, and each output node is associated with a class of data.

4 The SOM-PORTRAIT Method

In this section, we present our method, showing how the training set was built, describing our SOM-PORTRAIT method, and giving details of the model configuration and implementation characteristics.

4.1 Training Set Construction

The training set for SOM-PORTRAIT was built following exactly the same steps designed for the SVM-PORTRAIT method [3], which allowed us to compare the results obtained from both methods. The SOM-PORTRAIT training set consists of two sets, one composed by known protein-coding sequences (mRNAs) and the other formed from known non-coding sequences (ncRNAs).

The mRNAs set was primarily downloaded from the SwissProt database [2], version 50.8, in october 2006. Redundant sequences were eliminated using CD-HIT [15] with sequence similarity above 70% and running BLASTCLUST over the remaining sequences. The ORF prediction was done by ANGLE [23], and the obtained results were separated into three distinct files: a file containing nucleotide sequences with predicted ORFs, a file containing the corresponding amino acid sequences as predicted by ANGLE, and another file containing nucleotide sequences without predicted ORFs.

The ncRNAs set was also downloaded in october 2006 and contained sequences from NONCODE [16], Rfam [9] and RNAdb [21] databases, with redundant sequences eliminated using BLASTCLUST. The ANGLE prediction was done exactly in the same way as described for the mRNAs set.

4.2 The Method Description

The SOM-PORTRAIT method uses *ab initio* steps to predict ncRNA in a transcriptome. The method was originally conceived for two classes (coding RNA and non-coding RNA), but it could be easily modified to create more classes. We also worked with a three classes model, as discussed later.

The method used ANGLE to identify nucleotide sequences with and without predictable ORFs, but another ORF predictor could be used as well. Besides, attributes from the sequences with predicted ORFs were extracted separately from those without predicted ORFs. Sequence attributes must be converted to numerical data, that are the real input for the SOM-PORTRAIT. These attributes were chosen based on their hypothetical relevance to distinguish mRNAs from ncRNAs, and they are exactly the same as those adopted for the SVM-PORTRAIT experiments. We developed PERL scripts to extract, from each sequence, 111 variables, which were grouped into the following 7 main attributes, listed with the number of the variables generated by each attribute: nucleotide composition (84 variables), ORF length (4 variables), amino acid composition (20 variables), protein isoeletric point (1 variable), protein complexity (1 variable), mean protein hidropathy (1 variable), length (4 variables).

Note that attributes 2 to 6 refer only to sequences with predicted ORFs. Attributes 1 and 7 are extracted from both sequences with and without ORFs. So, the protein-dependent model consists of 111 variables comprised in 6 attributes, and the protein-independent model consists of 88 variables comprised in 2 attributes.

The extracting attributes step generates two input data files, each one containing numerical data relative to all the nucleotide sequences. The first file, containing sequences with predicted ORFs, is sent to prediction using a protein-dependent SOM (called **model.withorf.map**), and the second file, containing sequences without predicted ORFs, is sent to prediction using the protein-independent SOM (called **model.withoutorf.map**). Details for the two maps configuration are explained in the following section. Figure 2 shows the SOM-PORTRAIT workflow, together with the main input/output files.

4.3 Model Configuration and Implementation Details

We used the SOM-PAK [13] library (version 3.1) to configure the two models, one with ORFs and the other without ORFs. The machine used for training was a PC with a dual processor (Core 2 Duo 2.0 Ghz) and a 2,024 MHz RAM, executing Linux Ubuntu 8.04 (kernel 2.6.24-24-generic).

First, we configured a two classes model to classify the transcripts in coding RNA (mRNA) and non-coding RNA (ncRNA). Following, we configured a three classes model, creating an additional class labeled "undefined", trying to refine the classification of transcripts in another RNA class. For each of these experiments, two SOMs were configured, respectively, *model.withorf.map* and *protein-dependent*, for sequences with predicted ORFs, and *model.withoutorf.map* and *protein-independent*, for sequences without predicted ORFs.

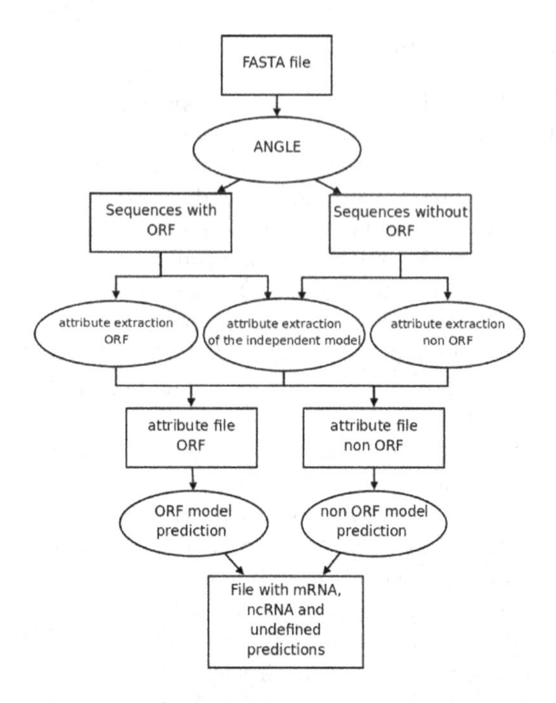

Fig. 2. The workflow of the SOM-PORTRAIT program

The execution of the SOM-PAK library created many files from the original training sets, and the desired attributes were extracted using a PERL script (*parameter_extractor.pl*). This script ran approximately 11 hours for the protein-dependent model and 3 hours for the protein-independent model, for both SOM models - with two and three classes. Also, both SOM-PORTRAIT models executed in less than 10 minutes of real CPU time, with time estimated using the *time* shell command, in the machines described in Section 4.3.

Two Classes. For the two classes experiment, the attribute files were further subdivided (Table 1). We note that in general a greater number of classes are used for SOM models. This experiment was developed in order to compare the results produced by SOM-PORTRAIT and the two classes SVM-PORTRAIT.

The configuration of the two classes SOM model was performed on three steps. First, the map was initialized as follows. SOM-PAK has a random initial distribution of the weights in the map. We chose a map with two nodes (representing the two classes), a rectangular topology and the gaussian function as the neighborhood function $h_j(t)$. The input data were the *.dat* files shown in Table 1. The second step was to organize the values of the *.dat* files between the two nodes initialized in the first step. We set the training rate α to 0.05, the radius for the initial neighborhood to 1, and the iteration value to 20,000 steps. These parameters were chosen following the recommendation for SOM training of SOM-PAK [13] and took about 10 minutes. On the third step, a smaller

Table 1. Each column shows, for each sequence file, respectively, its name, contents and number of sequences

Name	Purpose	Sequences
model.withorf.dat	data file for ORF map creation	20,000
model.withoutorf.dat	data file for non-ORF map creation	12,555
model.withorf.test	test file for quantization error estimate	76,827
model.withoutorf.test	test file for quantization error estimate	47,151

training rate $\alpha(0) = 0.02$ was adopted, the radius for the initial neighborhood was $h_j(0) = 1$, and the number of iterations t was fixed in $200,000$ steps.

The trained map was submitted to a large test file, containing all the attributes extracted from the training set sequences (*.test* files), to estimate the average quantization error (AQE). AQE is a statistical measure of the SOM training accuracy, and is calculated as the mean of the Euclidean distance $\|x_i(t) - v_c(t)\|$, where $v_c(t)$ is the winner node computed by SOM. In our experiments, the protein-dependent map AQE was 0.93, while the protein-independent map AQE was 0.64.

Three Classes. Then, we configured a three classes SOM model, performing training and configuration analogously to the two classes SOM-PORTRAIT (Table 1). In the first step, we chose a map with three nodes (representing the three classes), a rectangular topology and the gaussian function as the neighborhood function, analogously to the two classes model. The input data was the *.dat* files in Table 1. In the second step of organizing the map, we set the training rate α to 0.05, the radius for the initial neighborhood to 2, and the iteration value to $20,000$ steps. We used the same parameters adopted for the two classes experiment. For the last step, we used a smaller training rate of $\alpha(0) = 0.02$, the radius for the initial neighborhood was set to 1, and the number of iterations was fixed in $200,000$ steps. SOM-PAK estimated 0.79 for the protein-dependent map AQE, and 0.49 for the protein-independent map AQE.

5 A Case Study: The *Paracoccidioidis Brasiliensis* Fungus

The testing set used to validate the SOM-PORTRAIT method was the transcriptome of the *Paracoccidioidis brasiliensis* fungus, named *Pb* transcriptome, that has 6,022 Assembled ESTs [7]. A PERL script was developed to filter this data set, accepting a sequence if it had length with at least 80 nucleotides and at most 20% of characters different from A, C, G and T. This filter script discarded 9 sequences, from the 6,022 *Pb* Assembled ESTs, generating a final testing set containing 6,013 Assembled ESTs. The machines used for testing and validation were the same described in Section 4.3.

SOM-PORTRAIT workflow (Figure 2) was implemented in PERL. The testing set was executed by the program in 13 minutes of real CPU time.

A local version of the SVM-PORTRAIT was completely reimplemented. Training, testing and validation were done exactly as specified by Arrial and co-authors [3,4]. We used four machines for training and validation, all with processor Core 2 Duo 2.2 GHz, three with 2,024 Mhz RAM (Machines 1, 3 e 4), and one with 3,036 Mhz RAM (Machine 2). Machines 1 and 2 were used for training and validating the protein-dependent SVM model, while machines 3 and 4 were used for training and validating the protein-independent SVM model. For training and validation, the parameters and optimal values were adjusted following the specifications of the SVM-PORTRAIT. These two steps executed in approximately 40 hours for the protein-dependent model and 30 hours for the protein-independent model of real CPU time.

Now, we show two experiments with SOM-PORTRAIT, the first one with two classes and the second one with three classes, comparing the obtained results with the SVM-PORTRAIT output. Finally, we compare the results obtained by the two SOM-PORTRAIT models.

For these comparisons, we used as input the sequences of SVM-PORTRAIT presenting classification probability above 70%, which discarded 645 sequences, or 10,78% from the above 6,013 sequences. So, for the experiments, we considered a total of 5,365 Assembled ESTs from the *Pb* transcriptome.

5.1 Experiment 1: Two Classes

The same test steps done by SVM-PORTRAIT to classify the Assembled ESTs of the *Pb* transcriptome were repeated for the two classes SOM-PORTRAIT. The classification step executed 12 minutes of real CPU time. Table 2 shows that the percentages of coding sequences and ncRNAs found by SVM-PORTRAIT and the two classes SOM-PORTRAIT are very close. The third portion of this table shows a comparison between the classification as coding and non-coding sequences done by both methods.

Table 2. Comparisons between the two classes SOM-PORTRAIT and SVM-PORTRAIT. The percentage is relative to the 5,365 sequences of the *Pb* transcriptome.

	Sequences	**Percentage**
SVM-PORTRAIT		
Coding sequences	4,705	87.70%
ncRNAs	660	12.30%
SOM-PORTRAIT		
Coding sequences	4,689	87.40%
ncRNAs	676	12.60%
SOM-PORTRAIT × SVM-PORTRAIT		
coding/coding	4,337	80.83%
non-coding/non-coding	308	5.74%
coding/non-coding	368	6.86%
non-coding/coding	352	6.56%

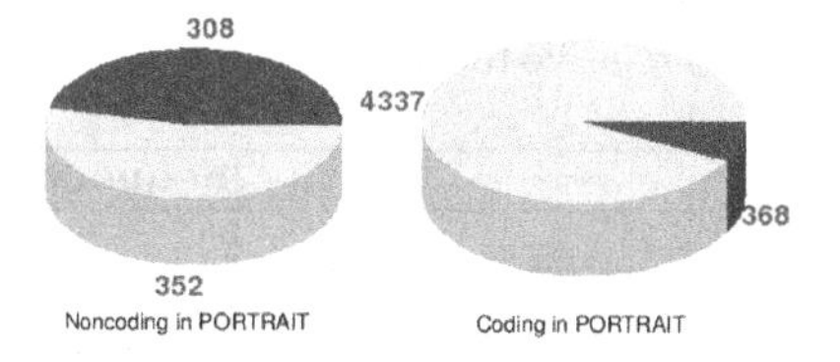

Fig. 3. Classifications produced for the Assembled ESTs of the *Pb* transcriptome by the SVM-PORTRAIT and the two classes SOM-PORTRAIT

Comparisons of Table 2 are shown in Figure 3, in which each circle represents the total of sequences classified as non-coding (left) and coding (right) by SVM-PORTRAIT, and inside each circle we represent the total of sequences classified by SOM-PORTRAIT as non-coding (black) and coding (light gray).

These results show that the two methods found very different classification for the input data, specially for non-coding sequences. This could be explained by the fact that the class of ncRNAs is very heterogeneous. SVM-PORTRAIT defined two classes, which means that the training step of this method included all ncRNAs in just one class. For the SOM-PORTRAIT model, we did not force the classification, that is, each class is built by the method taking the closest sequences. This analysis led us to develop another experiment including more classes in SOM-PORTRAIT, described in the next section. In fact, for the three classes SOM model, these results were improved, as we will see. Nonetheless, the non-supervised learning algorithm adopted in our SOM-PORTRAIT method reached a data classification accuracy comparable to that obtained by the SVM-PORTRAIT supervised learning method, with the clear advantages of reducing training time and needing less computational memory.

5.2 Experiment 2: Three Classes

The 5,365 sequences of the *Pb* transcriptome were submitted as input for a three classes SOM-PORTRAIT. The classification step executed in 13 minutes of real CPU time. Table 3 shows that the percentages of ncRNAs found by SVM-PORTRAIT and the three classes SOM-PORTRAIT remain very close.

Comparisons of Table 3 are shown in Figure 4, in which each circle represents the total of sequences classified as non-coding (left) and coding (right) by SVM-PORTRAIT, and inside each circle we represent the total of sequences classified by the three classes SOM-PORTRAIT as non-coding (dark gray), coding (light gray) and "undefined" (black).

We can note that the "undefined" class was composed most by the SVM-PORTRAIT non-coding class. The sequences classified by SOM-PORTRAIT as coding when compared to those found by the SVM-PORTRAIT non-coding class slightly shrink, indicating that the new "undefined" class grouped some of the sequences that were differently classified in the first experiment. This might indicate new possibilities for classifying, that could be done by the SOM model. Furthermore, it could be directly correlated to the great number of different

Table 3. Comparisons between the three classes SOM-PORTRAIT and SVM-PORTRAIT. The percentage is relative to the 5,365 sequences of the *Pb* transcriptome.

	Sequences	Percentage
SOM-PORTRAIT		
Coding sequences	4,357	81.21%
ncRNAs	676	12.60%
"Undefined´´ sequences class	332	6.19%
SVM-PORTRAIT × SOM-PORTRAIT		
coding/coding	4,325	80.62%
non-coding/non-coding	308	5.74%
coding/"undefined"	12	0.22%
non-coding/'undefined"	320	5.96%
coding/non-coding	368	6.86%
non-coding/coding	32	0.60%

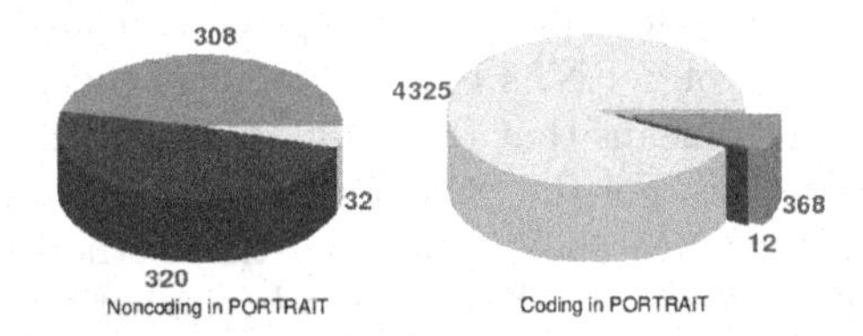

Fig. 4. Comparisons between the classification for the *Pb* transcriptome produced by the SVM-PORTRAIT and the three classes SOM-PORTRAIT

classifications of the non-coding sequences found by the two classes SOM-PORTRAIT with respect to the SVM-PORTRAIT, which indicates that we could create more classes in the SOM model to increase the classification accuracy.

5.3 Comparing Two and Three Classes SOM-PORTRAIT

Figure 5 shows a comparison between both SOM-PORTRAIT models, in which each circle represents the total of sequences classified as non-coding (left) and coding (right) by the two classes SOM-PORTRAIT, and the portions inside each circle shows the total of sequences classified as non-coding (dark gray), coding (light gray) and "undefined" (black) by the three classes SOM-PORTRAIT. Notice that the non-coding class remains almost the same on both models.

It is interesting to note that the non-coding class of SOM-PORTRAIT remains constant between two and three classes classifier, but a fraction of the coding sequences of the two classes SOM-PORTRAIT was transferred to the "undefined" class of the three classes classifier. But, this "undefined" fraction was classified as non-coding by the SVM-PORTRAIT. Therefore, it is tempting to explain this "undefined" class as a *bona fide* ncRNA quite close to coding RNA, that could be wrongly classified when the classification is restricted to two classes.

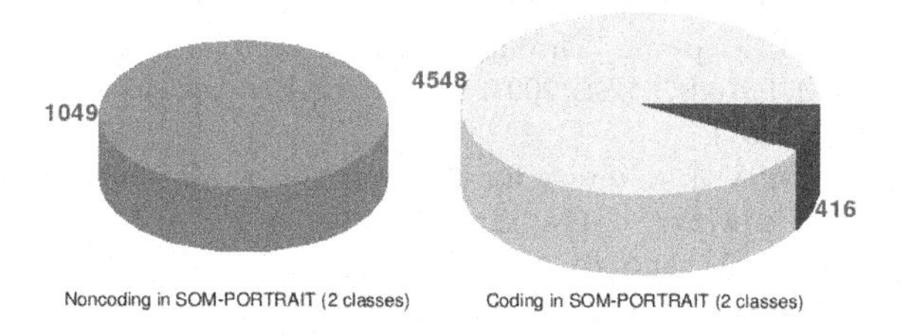

Fig. 5. Comparisons of the classifications for the 6,013 Assembled ESTs of the *Pb* transcriptome produced by two classes and three classes SOM-PORTRAIT

6 Conclusions and Future Work

In this work we proposed a method to identify non-coding RNAs using Self Organizing Maps (SOM), named SOM-PORTRAIT. We implemented the method and applied it to a data set containing Assembled ESTs of the *Paracoccidioides brasiliensis* fungus transcriptome. The obtained results were reliable, when compared to a method based on the SVM paradigm, noting that the time and memory requirements needed to our SOM-PORTRAIT is much less than those needed for methods based on the SVM paradigm, like PORTRAIT.

The following step is to assess the sensitivity and accuracy of our method, which could be done by applying SOM-PORTRAIT to randomly chosen known coding and non-coding RNAs and comparing the results to other ncRNA predictor methods. We also could test how much ANGLE affects the accuracy of the SOM-PORTRAIT method, removing ANGLE from the method and considering only the protein independent parameters to analyze how much the accuracy would be changed. Other interesting works are to develop WEB interfaces for the SOM-PORTRAIT allowing the user to select features to create classes according to his needs, to create more classes to see if specific sets of ncRNAs could be found, to use training pruners, and to include confidence level to the classification.

References

1. Altschul, S.F., Madden, T.L., Schäffer, A.A., Zhang, J., Zhang, Z., Miller, W., Lipman, D.J.: Gapped BLAST and PSI-BLAST: a new generation of protein database search programs. Nucleic Acids Res 25(17) (1997)
2. Apweiler, R., Bairoch, A., Wu, C.H., Barker, W.C., Boeckmann, B., Ferro, S., Gasteiger, E., Huang, H., Lopez, R., Magrane, M., Martin, M.J., Natale, D.A., O'Donovan, C., Redaschi, N., Yeh, L.S.: Uniprot: the universal protein knowledgebase. Nucleic Acids Res. 32, D115–D119 (2004)
3. Arrial, R.T.: Predicting noncoding RNAs in the transcriptome of the Paracoccidioides brasiliensis fungus using machine learning. Master's thesis, University of Brasilia (2008) (in Portuguese)

4. Arrial, R.T., Togawa, R.C., Brigido, M.M.: Outlining a Strategy for Screening Noncoding RNAs on a Transcriptome Through Support Vector Machines. In: Sagot, M.-F., Walter, M.E.M.T. (eds.) BSB 2007. LNCS (LNBI), vol. 4643, pp. 149–152. Springer, Heidelberg (2007)
5. Chang, C., Lin, C.: LIBSVM: a library for Support Vector Machines(2001), `http://www.csie.ntu.edu.tw/~cjlin/libsvm`
6. Eddy, S., `http://infernal.janelia.org/`
7. Felipe, M.S.S., co authors: Transcriptional profiles of the human pathogenic fungus Paracoccidioides brasiliensis in mycelium and yeast cells. Journal of Biological Chemistry (2005) doi:10.1074/jbc.M500625200
8. Griffiths-Jones, S.: Annotating Noncoding RNA Genes. Annu. Rev. Genomics Hum. Genet. 8, 279–298 (2007)
9. Griffiths-Jones, S., Moxon, S., Marshall, M., Khanna, A., Eddy, S.R., Bateman, A.: Rfam: annotating non-coding RNAs in complete genomes. Nucleic Acids Research 33, D121–D124 (2005), `http://www.sanger.ac.uk/Software/Rfam/`
10. Haykin, S.: Neural Networks. Macmillan College Publishing Company, New York (1994)
11. Hofacker, I.L., Fekete, M., Stadler, P.F.: Secondary Structure Prediction for Aligned RNA Sequences. Journal of Molecular Biology 319(5) (2002)
12. Kohonen, T.: Self-Organization and Associative Memory. Springer, New York (1998)
13. Kohonen, T., Hynninen, J., Kangas, J., Laaksonen, J.: SOM_PAK: The Self-Organizing Map Program Package. Technical report, Helsinki University of Technology, Espoo, Finland (1996)
14. Kong, L., Zhang, Y., Ye, Z.-Q., Liu, X.-O., Zhao, S.-O., Wei, L., Gao, G.: CPC: assess the protein-coding potential of transcripts using sequence features and support vector machine. Nucleic Acids Res. 35, 345–349 (2007)
15. Li, W., Godzik, A.: CD-HIT: a fast program for clustering and comparing large sets of protein or nucleotide sequences. Bioinformatics, 22(13) (2006)
16. Liu, C., Bai, B., Skogerbo, G., Cai, L., Deng, W., Zhang, Y., Bu, D., Zhao, Y., Chen, R.: NONCODE: an integrated knowledge database of non-coding RNAs. Nucleic Acids Research 33, D112–D115 (2005), `http://www.noncode.org/`
17. Liu, J., Gough, J., Rost, B.: Distinguishing protein-coding from non-coding RNAs through Support Vector Machines. PLoS Genet. 2(4) (April 2006)
18. Liu, J., Gough, J., Rost, B.: Distinguishing protein-coding from non-coding rnas through support vector machines. PLoS Genet. 2(e), 29–36 (2006)
19. Mattick, J.S.: Non coding RNAs: the architects of eukaryotic complexity. EMBO reports 2, 986–990 (2001)
20. Mount, S.M., Gotea, V., Lin, C.-F., Hernandez, K., Makalowski, W.: Spliceosomal small nuclear RNA genes in eleven insect genomes. RNA 13(1), 5–14 (2007)
21. Pang, K.C., Stephen, S., Engstrom, P.G., Tajul-Arifin, K., Chen, W., Wahlestedt, C., Lenhard, B., Hayashizaki, Y., Mattick, J.S.: RNAdb - a comprehensive mammalian noncoding RNA database. Nucleic Acids Research 33, D125–D130 (2005), `http://jsm-research.imb.uq.edu.au/rnadb/About/default.aspx`
22. Rumelhart, D.E., Zipser, D.: Feature discovery by competitive learning. Cognitive Science 9 (1985)

23. Shimizu, K., Adachi, J., Muraoka, Y.: ANGLE: a sequencing errors resistant program for predicting protein coding regions in unfinished cDNA. Journal of Bioinformatics and Computational Biology 4(3), 649–664 (2006)
24. Washietl, S., Hofacker, I.L., Lukasser, M., Stadler, P.F., Hüttenhofer, A.: Mapping of conserved RNA secondary structures predicts thousands of functional noncoding RNAs in the human genome. Nat. Biotechnol. 22, 1383–1390 (2005)
25. Watson, J.D., Crick, F.H.C.: A structure for deoxyribose nucleic acid. Nature, 171 (1953)
26. Zucker, M., Matthews, D.H., Turner, D.H.: Algorithms and thermodynamics for RNA secondary structure prediction: A practical guide. In: RNA Biochemistry and Biotechnology, NATO ASI Series, pp. 11–43. Kluwer Academic Publishers, Dordrecht (1999)

Automatic Classification of Enzyme Family in Protein Annotation

Cássia T. dos Santos[1], Ana L.C. Bazzan[2], and Ney Lemke[3]

[1] Departamento de Informática, Universidade de Évora, Portugal
cassia.ts@gmail.com
[2] Instituto de Informática / PPGC, Universidade Federal do Rio Grande do Sul
C. P. 15064, 91.501-970, Porto Alegre, RS, Brazil
bazzan@inf.ufrgs.br
[3] Dep. de Física e Biofísica, Instituto de Biociências, UNESP
C.P. 510, 18618-000, Botucatu, SP, Brazil
lemke@ibb.unesp.br

Abstract. Most of the tasks in genome annotation can be at least partially automated. Since this annotation is time-consuming, facilitating some parts of the process – thus freeing the specialist to carry out more valuable tasks – has been the motivation of many tools and annotation environments. In particular, annotation of protein function can benefit from knowledge about enzymatic processes. The use of sequence homology alone is not a good approach to derive this knowledge when there are only a few homologues of the sequence to be annotated. The alternative is to use motifs. This paper uses a symbolic machine learning approach to derive rules for the classification of enzymes according to the Enzyme Commission (EC). Our results show that, for the top class, the average global classification error is 3.13%. Our technique also produces a set of rules relating structural to functional information, which is important to understand the protein tridimensional structure and determine its biological function.

1 Introduction

One of the major challenges of post-genomic bioinformatics is to automate the annotation of a large number of available protein sequences. Ultimately, this annotation means a characterization of a given protein structure, function, cellular metabolism, and its interactions with other proteins.

The annotation process assigns putative functions to a given protein. Once this process is complete, we can use this information to extract relevant information about the organism metabolism. Besides, intermediate stages of annotation process include annotating or filling certain information in databases.

Protein function annotation includes classification of protein sequences according to their biological function and is an important task in bioinformatics. Of course, ideally, manually curated databases are preferred over automated processing. Issues such as the quality of data and compliance with standards could

K.S. Guimarães, A. Panchenko, T.M. Przytycka (Eds.): BSB 2009, LNBI 5676, pp. 86–96, 2009.

then be carefully analyzed. However, this approach is expensive, if feasible in the first place. With the large number of protein sequences entering into databases, it is desirable to annotate new incoming sequences by means of an automated method.

Approaches based on data mining and machine learning have been largely used, which explore functional annotation by using learning methods including decision trees and instance-based learning [1,4,7], neural networks [9], and support vector machines [3].

The most common way of annotating the function of a protein is to use sequence similarity (homology). Methods such as Smith-Waterman algorithm and BLAST have been largely used for measuring sequence similarity. However, the function of a protein with very small homology is difficult to assess using these methods, even using PSI-blast due to the low significance of the first hit. The same problem may arise for homologous proteins of different functions if one is only recently discovered and the other is the only known protein of similar sequence. This way, it is desirable to explore methods that are not based on sequence similarity. Protein sequence motifs can be seen an alternative approach.

Motifs can reveal important clues to a protein's role even if it is not globally similar to any known protein. The motifs for most important regions such as catalytic sites, binding sites, and protein-protein interaction sites are conserved over taxonomic distances and evolutionary time than are the sequences of the proteins themselves.

In general, a pattern of motifs is required to classify a protein into a certain family of proteins [2]. This is a very valuable source of information as gives a good idea of the variation of functions within the family, and hence a good indicative of how much functional annotation can safely be transferred. In summary, motifs are highly discriminative features for predicting the function of a protein.

Our approach uses symbolic machine learning techniques to induce rules to annotate the functional families using motifs. In this paper we restrict ourselves to the specific case of enzymes, since it is one of the most relevant classes. We avoid the use of neural network or support vector machine because, since the rules generated will be evaluated by a human expert, it is important that s/he can understand how the rules were generated. The symbolic methods are more elucidative regarding this issue.

Specifically, we are interested in classifying enzymes into functional classes according to the Enzyme Commission (EC) top level hierarchy. Such classifier will be then applied to unannotated proteins. This is indeed important because knowing to which family or subfamily an enzyme belongs may help in the determination of its catalytic mechanism and specificity, giving clues to the relevant biological function. Notice that this information is not directly available in databases such as Interpro. Although it may appear in SWISS-PROT (DE line), since it is manually curated, the EC classification is not always available. The present paper describes exactly a way to fill this part of the DE line in an automatic way, thus being suitable to be inserted in TrEMBL.

2 Enzyme Classification

Enzymes are classified using the so-called EC (Enzyme Commission) codes. A given class represents the function of an enzyme and is specified by a number composed by four digits (e.g. 1.1.1.1 for alcohol dehydrogenase). The first digit indicates the general type of chemical reaction catalyzed by the enzyme.

The others three digits have meanings that are particular to each class. For instance, in the oxidoreductases (EC number starting with 1), which involve reactions in which hydrogen or oxygen atoms or electrons are transferred between molecules, the second digit specifies the chemical group of the donor molecule. The third digit specifies the acceptor, and the fourth digit specifies the substrate.

Considering that a particular protein can have several enzymatic activities, it can be classified into more than one EC classes. These proteins are considered multi-functional.

3 Databases

In bioinformatics, electronic databases are necessary due to the explosion and distribution of data related to the various genome projects. Here, the databases related to the present paper are briefly presented.

3.1 SWISS-PROT/Uniprot

SWISS-PROT (Uniprot)[1] [6] is a curated database which provides a high level of annotation of each sequence, including: a description of the function of a protein, its domain structure, post-translational modifications, variants, etc. SWISS-PROT has also extensive links to other databases, such as the databases of motifs and enzymes. An example entry to SWISS-PROT can be seen in Figure 1. In this particular case the entry is an enzyme. Parts of the entry were deleted to render the figure smaller. The first lines are the protein ID and accession number (AC). For our purposes the "DE" (description) and the "DR" (cross reference to other databases) lines are the most important. In this specific case, DE shows the name of the enzyme and its EC code. The DR lines show how this enzyme entry in SWISS-PROT is related to other databases, in particular to motifs databases: it belongs to the InterPro family IPR010452 and to the Pfam class PF06315. This is important because some of these cross references will be used as attributes in the rule induction process, as it will be detailed.

3.2 Databases of Motifs

Since there are many motifs' recognition methods to address different sequence analysis problems, different databases of motifs exist, including those that contain relatively short motifs (e.g., PROSITE[2]); groups of motifs referred to as

[1] `http://www.expasy.ch/sprot/`
[2] `http://www.expasy.ch/prosite`

```
ID  ACEK_PSEAE     STANDARD;     PRT;  577 AA.
AC  Q9I3W8;
[...]
DE  Isocitrate dehydrogenase kinase/phosphatase (EC 2.7.1.116)
DE  (EC 3.1.3.-) (IDH kinase/phosphatase) (IDHK/P).
GN  Name=aceK; OrderedLocusNames=PA1376;
OS  Pseudomonas aeruginosa.
OC  Bacteria; Proteobacteria; Gammaproteobacteria; Pseudomonadales;
[...]
CC  -!- FUNCTION: Bifunctional enzyme which can phosphorylate or
CC      dephosphorylate isocitrate dehydrogenase (IDH) on a specific
CC      serine residue. This is a regulatory mechanism which enables
[...]
CC  This Swiss-Prot entry is copyright. It is produced through a collaboration
CC  between the Swiss Institute of Bioinformatics and the EMBL outstation -
CC  the European Bioinformatics Institute. There are no restrictions on its
CC  use as long as its content is in no way modified and this statement is not
CC  removed.
[...]
DR  InterPro; IPR010452; AceK.
DR  Pfam; PF06315; AceK; 1.
DR  PIRSF; PIRSF000719; AceK; 1.
KW  ATP-binding; Complete proteome; Glyoxylate bypass; Hydrolase; Kinase;
KW  Multifunctional enzyme; Protein phosphatase; Transferase;
KW  Tricarboxylic acid cycle.
[...]
SQ  SEQUENCE   577 AA;  66761 MW;  B3E8A69981D5DC60 CRC64;
   MVQSAPASEI AALILRGFDD YREQFREITD GARARFEQAQ WQEAQ [...]
```

Fig. 1. SWISS-PROT entry for protein Q9I3W8

fingerprints (e.g. PRINTS[3]); or sequence patterns, often based on position-specific scoring matrices or Hidden Markov Models generated from multiple sequence alignments (e.g. Pfam[4]).

Interpro[5] integrates the most commonly used motifs databases, providing a layer on the top of them. Thus, a unique, non-redundant characterization of a given protein family, domain or functional site is created. InterPro receives the signatures from the member databases. The motifs are grouped manually into families (there are some automatic methods for producing the matches and overlap files, but the ultimate decision for grouping and for determining relationships comes from a biologist).

[3] http://umber.sbs.man.ac.uk/dbbrowser/PRINTS/

[4] http://www.sanger.ac.uk/Pfam/

[5] http://www.ebi.ac.uk/interpro/

3.3 Databases of Enzymes

Enzyme databases contain data about enzyme nomenclature, classification, functional characterization, and cross-references to others database, such as SWISS-PROT.

ENZYME[6] is a repository of information relative to the nomenclature of enzymes, based on the recommendations of the IUBMB. For each entry, it is possible to obtain a list of all corresponding proteins in SWISS-PROT database. This way, is possible to obtain all proteins classified as enzymes. Moreover, SWISS-PROT entries are also linked to the ENZYME database. As said, the line "DE" in the SWISS-PROT entry contains the EC numbers associated to the corresponding protein. As Figure 1, the protein Q9I3W8 has two ECs numbers associated – 2.7.1.116 ([Isocitrate dehydrogenase (NADP+)] kinase) and 3.1.3.-. (Hydrolases acting on ester bonds) – which are described in the "DE" line. These ECs are links to the ENZYME database. Linking these two databases is an efficient way to obtain the list of proteins annotated with the EC classes.

Other important database is BRENDA (BRaunschweig ENzyme DAtabase)[7], a comprehensive collection of enzyme and metabolic information. It includes biochemical and molecular information on classification and nomenclature, reaction and specificity, functional parameters, occurrence, enzyme structure, application, engineering, stability, disease, isolation and preparation, links and literature references.

4 Methods and Results

4.1 Data Processing

Our approach aims at the prediction of protein function, by classifying proteins into functional families, according to the enzymatic function classification. Initially, we focused on the top level of the EC hierarchy, and later extended the classification to the second level. The C4.5 [8] algorithm was used to generate the decision tree.

Regarding training data, we used all proteins from the SWISS-PROT database in which the line "DE" is annotated with EC numbers. The first digit of the EC numbers appearing in this field in the corresponding SWISS-PROT entries were used. The data comes from a local version of the SWISS-PROT database. Using it, around 60,000 proteins annotated with EC numbers were found.

The attributes which appear in this data are accession numbers related to InterPro domain classifications. These appear in SWISS-PROT as a cross-referenced database (see Figure 1). This way, for each SWISS-PROT entry, the InterPro accession numbers and corresponding EC numbers, considering only the first and second digits, were extracted (i.e, each entry corresponds to one learning instance). These InterPro accession numbers were used as attributes by the learning algorithm. The attributes which appear in all examples related to one specific class

[6] http://au.expasy.org/enzyme/

[7] http://www.brenda.uni-koeln.de/

Table 1. EC classes and number of attributes

EC	Functional Class	Attributes	EC	Functional Class	Attributes
1	Oxidoreductase	715	2	Transferase	1309
3	Hydrolase	1269	4	Lyase	391
5	Isomerase	231	6	Ligase	319

```
EC3,n.

IPR005965: y,n.
IPR001926: y,n.
IPR003029: y,n.
IPR001247: y,n.
IPR002381: y,n.
IPR001041: y,n.
[...]
```

Fig. 2. Input data for EC 3 class (partially shown)

are used to characterize it. The exact number of attributes found for each EC class appears in Table 1; the number of instances is, as mentioned, around 60 thousand.

This data is organized in input files, one for each EC class, respecting the specific syntax of C4.5. Basically, two files (.names and .data) are used. The names file is formed by lines defining the names of classes, attributes and attribute values. The .data file contains one line for each protein, where each line contains the values of the attributes, separated by commas. Figure 2 shows (partially) an example of names file corresponding to EC 3 class.

Some proteins are multi-functional and belong to more than one EC class. For these proteins, more than one instance is created. For instance, the protein Q9I3W8 (Figure 1) is considered a positive instance to the EC 2 and EC 3 classes. This way, a total of 60488 instances were used.

4.2 Methods

Using the input data, C4.5 induces rules for a given target class. Figure 3 shows a partial output of the C4.5. In this specific case, the rule refers to the class "EC3". The "|" sign indicates an "AND". Thus, the rule basically suggests the annotation of the class "EC3" for a given enzyme if it belongs to the IPR001247 and if it does not belong to the IPR002381 families of proteins. If the protein does not belong to the IPR001247, then, the rules goes on asking if it belongs to the family IPR003029. If the answer is yes, then that class should be annotated, otherwise continue reading the rule and performing the nested tests. At the end, either some "IF" part is satisfied, or the class should not be annotated. The protein Q9I3W8 (Figure 1) is correctly classified into the EC3 class, since it is related to the IPR010452 (see "DE" line in Figure 1).

```
IPR001247 = y:
|   IPR002381 = y: n (36.0/1.0)
|   IPR002381 = n: EC3 (52.0/10.0)
IPR001247 = n:
|   IPR003029 = y: EC3 (27.0/12.0)
|   IPR003029 = n:
|   |   IPR011257 = y: EC3 (32.0/16.0)
|   |   IPR011257 = n:
|   |   |   IPR010452 = y: EC3 (26.0/13.0)
|   |   |   IPR010452 = n:[S9]
```

Fig. 3. Output of C4.5 for class EC 3 (partially shown)

Once the rules were generated, we have proceeded to the evaluation of the quality of these rules. We use the standard evaluation of the training data, mainly based on cross validation (CV), accuracy and confidence. Regarding the former, n-fold CV was used meaning that $1/n$ was used as test, n times. For accuracy we use $acc = \frac{TP+TN}{TP+FP+TN+FN}$ and for confidence: $c = \frac{p+\frac{z^2}{2n}-z\sqrt{\frac{p}{n}-\frac{p^2}{n}+\frac{z^2}{4n^2}}}{1+\frac{z^2}{n}}$, where $z = 1.96$ (for 95% of significance), $n = TP + FP$, and $p = \frac{TP}{n}$.

4.3 Results and Discussion

In principle, it is desirable that the error rate is as small as possible. The problem is that we have to analyze two classes: the negative (training instances which induce a rule indicating when a EC class should not be annotated); and the positive one (i.e. instances which indicate when an EC class should be annotated).

Table 2, column 2, shows the results regarding the global classification error for each EC class, performing the 10-fold CV procedure. The average error is 3.13% and 190 test instances had not been correctly classified (an average of 6048 test instances were used).

The classes EC 5 and EC 6 which have the lower number of attributes and training instances (Table 1), yield the best classification rates in the test data: 1% and 0.8% respectively. On other hand, the class EC 2 has the highest error (6.1%), having the highest number of attributes. This error rate (6.1%) is considered by specialists acceptable and a quite good classification rate.

If we observe Table 2, columns 4 and 7, we see that the error rate in the non class is lower than in the positive class, for all EC classes. The number of negative instances (counter examples) is higher than the positive instances (examples). This occurs because the negative instances of one class are composed by the positive instances of the others five classes.

Finally, regarding accuracy and confidence, for the former a figure of 96.58% was obtained. Regarding the class and non-class, the confidence is approximately 0.97 and 0.96 respectively.

Table 2. Results of the classification using 10-fold-cross-validation

EC	Global Error (%)	Class			Non Class		
		Instances	Error (%)	Conf.	Instances	Error (%)	Conf.
1	209.90 (3.50)	1160.3	198.00 (17.06)	0.98	4888.5	12.00 (0.25)	0.96
2	371.10 (6.10)	1958.4	342.20 (17.47)	0.98	4090.4	28.10 (0.69)	0.91
3	340.90 (5.60)	1485.7	305.10 (20.54)	0.96	4563.1	35.20 (0.77)	0.93
4	107.70 (1.80)	594.8	101.70 (17.10)	0.97	5454	5.80 (0.11)	0.98
5	60.90 (1.00)	283.6	54.20 (19.11)	0.94	5765.2	7.30 (0.13)	0.99
6	49.90 (0.80)	566	45.90 (8.11)	0.98	5482.8	5.10 (0.09)	0.99
Average	190.07 (3.13)	1008.13	174.52 (16.57)	0.97	5040.67	15.58 (0.34)	0.96

Table 3. 10-fold-cross-validation: global error (EC 6 second level)

EC	Global Error (%)
6.1	12.6 (2.2)
6.2	4.6 (0.8)
6.3	23.8 (4.2)
6.4	2.8 (0.5)
6.5	2.9 (0.5)
6.6	0.9 (0.2)
Average	7.9 (1.4)

The good results achieved with the first level classifiers have motivated us to extend the classification to the second level of the EC hierarchy. Thus, experiments with the six subclasses of the EC 6 class were performed. All proteins from the SWISS-PROT database which are annotated with "EC 6.x" numbers were used. 5640 proteins were found.

Table 3 shows the results regarding the global classification error for each EC 6 subclass, performing the 10-fold-cross-validation. The average error was 1.4% and only 7.9 test instances were not correctly classified. An average accuracy (*acc*) of 98.6% was obtained. Regarding the class and non-class, the average confidences were approximately 0.73 and 0.99, respectively.

The EC 6.6 subclass had the best classification rate, 0.2%, and the class EC 3 had the highest error (4.2%). These classification rates are better than those obtained using data from EC main classes. This is explained by the fact that data from EC 6 subclasses possesses smaller number of attributes and instances. The learning algorithms works better in such conditions, performing good generalizations on test data.

5 Related Work

Machine learning techniques have been largely used in bioinformatics and medical research to predict patterns, regularities, and to classify noisy, unstructured, as well as high-dimensional, class-unbalanced, heterogeneous data sets. In [7], the C4.5 algorithm was used to generate rules which were applied for automated annotation of *Keywords* in SWISS-PROT. Such rules can then be applied to unannotated protein sequences.

We skip the details and describe here just some issues relevant to the approach in this paper. In short, the authors have developed a method to automate the process of annotation regarding Keywords, based on the C4.5 algorithm. This algorithm works on training data, which, in this case, are previously annotated keywords regarding proteins. Such data comprises mainly taxonomy entries, InterPro classification, and Pfam and PROSITE patterns, which appear in SWISS-PROT as cross-referenced databases. Using these cross-references attributes, C4.5 derives a classification for a target class. Since dealing with the whole data in SWISS-PROT at once would be prohibitive, the authors divided it in protein groups according to the InterPro classification. Then each group was submitted to an implementation of C4.5. Rules were generated and a confidence factor for each rule was calculated.

One issue with this approach is that it uses too many attributes to train the model to induce rules for the automated annotation. To tackle this, in [1] the training instances were restricted to proteins related to a given family of organisms (in the case of this paper in particular, the *Mycoplasmataceae*). By doing this the performance of the algorithm has improved, both in terms of time and confidence: between 60% and 75% of the given keywords were correctly predicted. Besides, a comparison of the performance of symbolic and non-symbolic machine learning techniques was carry out. Regarding the non-symbolic approach, using an artificial neural network (ANN) with the backpropagation algorithm, the resulted classifier was more compact than the symbolic classifier. However, the ANN model consider all data at once, leading to a slightly worse performance than the symbolic approach.

Regarding this, similar approaches are reported. In [4] discrete naive Bayes, decision trees, and instance-based learning methods were used to induce classifiers that predict the EC class of an enzyme from features extracted from its primary sequence (such as pI, molecular weight, and amino-acid composition). Data derived from the Protein Data Bank (PDB) and from SWISS-PROT databases was used. The method predicted the first EC class of an enzyme with 74% accuracy and the second EC number of an enzyme with 68% accuracy. In 1997 – when the these results were obtained, a small number of entries in the SWISS-PROT was annotated as enzymes (approximately 15.000). Today this number is much higher (over 60.000). Thus, is it an open question whether these figures for accuracy could be repeated.

The use of support vector machine (SVM) for classification of enzymes into functional families is presented in [3]. SVM classification system for each family is trained from representative enzymes of that family and seed proteins of Pfam curated protein families. The classification accuracy for enzymes from 46 families and for non-enzymes is in the range of 50.0% to 95.7% and 79.0% to 100% respectively. The results show the potential of the approach for enzyme family classification and for facilitating protein function prediction.

Similarly to ours, the approach presented by [2] uses motif composition for annotation of enzymes, focusing the oxidoreductases class only. However, motifs were used to define a similarity measure or kernel function used with an kernel based classifier, namely SVM. Protein sequences annotated with EC numbers from the SWISS-PROT database were used as training data. The good results obtained show that most classes of enzymes can be classified using a handful of motifs, yielding accurate and interpretable classifiers.

The use of neural networks to classify enzymes found in the PDB is proposed by [9]. The main goal is to infer the function of an enzyme by analyzing its structural similarity to a given family of enzymes. The codification scheme converts the primary structure of enzymes into a real-valued vector.

This last two works for classification of *enzymes* ([2] and [9]) use non-symbolic learning methods and certain features extracted from primary sequences (i.e, pI, molecular weight, and amino-acid composition). The differences between our and those approaches are twofold. First, we adopted symbolic machine learning algorithms, considering that the classifiers will be evaluated/used by human experts and these algorithms generate representations more easy to interpret. Second, we use motifs to characterize the EC classes. Motifs are better than features extracted from primary sequences because a motif compactly captures the features from a sequence that are essential for the function of the protein; the catalytic site of a protein might be composed of several regions that are not contiguous in sequence; and motifs can offer greater interpretability.

6 Conclusion and Future Work

We have presented a prediction method for classification of enzymes from sequence motifs. The method does not make use of sequence similarity. This approach yielded accurate and interpretable classifiers. These classifiers present good classification rates on the test data, i.e, an average accuracy of 96.85% for the top level. In a specific case, the EC 6 class obtained the best results, with a lower classification error (0.80%) and only 49 enzymes were not classified correctly (considering 6049 proteins on test data). Regarding the EC 6 subclasses, the classifiers had an average accuracy of 98.6%.

In the future we intend to work on issues such as the extension of the classification to the others levels of the EC hierarchy, i.e, third and fourth digits of the EC numbers for all main classes. Also, we want to reduce the number of motifs, while maintaining classification confidence, as well as to use others motifs database, such as Pfam and Prosite. Moreover, we plan to use agents that apply different methods to feature selection and negotiate the better features representing the training data as in [5]. The motivation for this is that learning agents can be responsible for applying different machine learning algorithms (different learning algorithms applied to the same data set hardly generate the same results), using subsets of the data, cooperating to generate a global and accurate classifier.

References

1. Bazzan, A.L.C., da Silva, S.C., Engel, P.M., Schroeder, L.F.: Automatic annotation of keywords for proteins related to mycoplasmataceae using machine learning techniques. Bioinformatics 18(S2), S1–S9 (2002)
2. BenHur, A., Brutlag, D.: Sequence motifs: highly predictive features of protein function. In: Feature extraction, foundations and applications, pp. 625–643. Springer, Heidelberg (2005)
3. Cai, C., Han, L., Ji, Z., Chen, Y.: Enzyme family classification by support vector machines. Proteins: Structure, Function, and Bioinformatics 55(1), 66–76 (2004)
4. des Jardins, M., Karp, P., Krummenacker, M., Lee, T., Ouzounis, C.: Prediction of enzyme classification from protein sequence without the use of sequence similarity. In: Proceedings of the International Conference on Intelligent Systems Molecular Biology, pp. 92–99 (1997)
5. dos Santos, C.T., Bazzan, A.L.C.: Integrating knowledge through cooperative negotiation – A case study in bioinformatics. In: Gorodetsky, V., Liu, J., Skormin, V.A. (eds.) AIS-ADM 2005. LNCS, vol. 3505, pp. 277–288. Springer, Heidelberg (2005)
6. Gasteiger, E., Jung, E., Bairoch, A.: Swiss-prot: Connecting biological knowledge via a protein database. Curr. Issues Mol. Biol. 3, 47–55 (2001)
7. Kretschmann, E., Fleischmann, W., Apweiler, R.: Automatic rule generation for protein annotation with the C4. 5 data mining algorithm applied on SWISS-PROT. Bioinformatics 17, 920–926 (2001)
8. Quinlan, J.R.: C4.5: Programs for Machine Learning. Morgan Kaufmann, San Francisco (1993)
9. Weinert, W., Lopes, H.: Neural networks for protein classification. Applied Bioinformatics 3(1), 41–48 (2004)

Representations for Evolutionary Algorithms Applied to Protein Structure Prediction Problem Using HP Model

Paulo H.R. Gabriel and Alexandre C.B. Delbem

Depto. Sistemas de Computação
ICMC/USP - Caixa Postal 668
13560-970 - São Carlos/SP, Brazil
{phrg,acbd}@icmc.usp.br

Abstract. Protein structure prediction (PSP) is a computational complex problem. To work with large proteins, simple protein models have been employed to represent the conformation, and evolutionary algorithms (EAs) are usually used to search for adequate solutions. However, the generation of unfeasible conformations may decrease the EA performance. For this reason, this paper presents two alternative representations that reduce the number of improper structures, improving the search process. Both representations have been investigated in terms of initial population in order to start the evolutionary process with promising regions. The results have shown a significant improvement in the fitness values (or, in other words, in solution quality).

Keywords: Protein Structure Prediction, Representation of Evolutionary Algorithms, HP Model, Cubic Lattice.

1 Introduction

Protein structure prediction (PSP) problem consists in predicting the three-dimensional structure of a protein. This problem is one of the main research areas in computational molecular biology, for three reasons [1]:

1. The three-dimensional shape of a protein is related to its function;
2. The fact that a protein can be made by 20 different types of amino acids makes the resulting three-dimensional structure very complex;
3. No simple and accurate method for determining the three-dimensional structure is known.

In order to deal with these difficulties, several heuristic-based methods have been investigated. They include evolutionary algorithms (EAs) [2], which have been used due to their flexibility to solve problems from different fields (Nature Sciences, Engineering, Computer Sciences, and others) with relatively few adaptations in the EA basic code [2].

K.S. Guimarães, A. Panchenko, T.M. Przytycka (Eds.): BSB 2009, LNBI 5676, pp. 97–108, 2009.

Moreover, to model the solutions (the predicted structures) simple discrete models have been employed. One of the best-studied models is the hydrophobic–hydrophilic model [3] (so-called HP model), where the hydrophobic interactions are considered to be the main force in the folding process, and the folding space is modeled by a three-dimensional lattice.

EAs have been applied to PSP using the HP model [4,5,6,8,9]. In order to provide only feasible solutions, EAs analyze all generated structures and assign a penalty to each infeasible conformation during the evolutionary process. However, the addition of penalization has a bias towards decreasing the EA performance [10], since all amino acid interactions are evaluated more than once.

In this paper we have investigated representations of EAs using HP model that avoid the generation of unfeasible solutions, in order to simplify the search process. Moreover, representation of EAs is a growing research area, and has received attention in the literature [11]. We have proposed two alternative representations, named *direct representation* and *conformation matrix*. The first encodes the solutions in three arrays to generate the spatial positions. The second uses a statical three-dimensional matrix which addresses the spatial position of the amino acids and eliminates all improper structures, i.e., solutions in which there are amino acids collisions.

These proposed techniques were evaluated in terms of initial population, using examples from the literature [6,8]. Results have shown that EAs with an appropriated representation can find more adequate individuals and, consequently, start the evolutionary process with more promising regions [12,13] of the search space.

This paper is organized as follows: Section 2 provides a background about the HP model considered in this study; Section 3 presents concepts of EAs and describes the proposed representations; in Section 4, we compare the proposed representations with the representation used in a standard EA in the initial population and, finally, Section 5 presents some conclusions and outlines future projects.

2 HP Model for Protein Structure Prediction

Proteins are macromolecules built from basic units, named amino acids. There are twenty different types of amino acids that can be connected to their neighbors in a sequence by *peptide bonds*. Despite the rigidness of this bond, the atomic links can rotate, guiding the protein folding. However, simulating all the folding process is computationally expensive, since several physical and chemical factors can affect such rotations. For this reason, simplified models are employed.

One of the most studied simplified models is the *hydrophobic–hydrophilic model* (HP model) proposed by Dill [3]. In this model, each amino acid is classified into two classes: hydrophobic or non-polar (`H`), and hydrophilic or polar (`P`), according to their interaction with water molecules. In addition, the sequence is assumed to be embedded in a lattice, used to discretize the space of conformations. The simplest topologies of lattice are the square lattice (shown

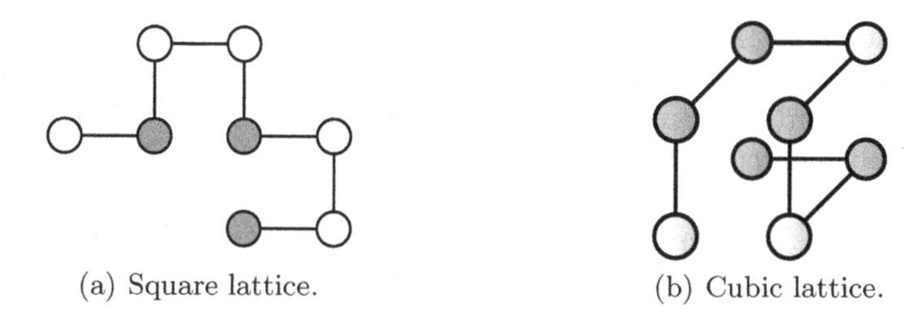

(a) Square lattice. (b) Cubic lattice.

Fig. 1. Example of conformation in square lattice (a) and cubic lattice (b) under the HP model. Gray balls represent the `H` amino acids and white balls represent the `P` ones.

in Fig. 1(a)) for two dimensions, and the cubic lattice (shown in Fig. 1(b)), for three dimensions. Other representations of structures can be found in [14].

Each pair of amino acids can be classified as *connected* or *neighbors*. Two amino acids from positions i and j in a sequence are connected if, and only if, $j = i+1$ or $j = i-1$. Notice that the number of connected amino acids is fixed and independent of conformation of the sequence. On the other hand, two amino acids in positions i and j are topological neighbors if they are not connected and the Euclidean distance between i and j is 1.

The native state of a protein is a low-energy conformation. Thus, each pair of *hydrophobic* neighbors (pair `HH`) in c contributes with a contact free energy -1; any other topological contact (`HP` or `PP`) does not contribute to the total free energy. Then, the number of `HH` contacts is maximized in the native state.

Formally, the free energy E is given by

$$E = -1 \sum_{1 \leq i+1 < j \leq n} B_{i,j}\delta(r_i, r_j),$$

where

$$B_{i,j} = \begin{cases} 1 \text{ if amino acids } i \text{ and } j \text{ are } H \text{ type;} \\ 0 \text{ otherwise;} \end{cases}$$

$$\delta(r_i, r_j) = \begin{cases} 1 \text{ if amino acids } r_i \text{ and } r_j \text{ are not conected;} \\ 0 \text{ otherwise.} \end{cases}$$

Despite the apparent simplicity of the model, finding the globally optimal conformation under the HP model is an $\mathcal{NP}$-complete problem [15,16], justifying the use of heuristic based-techniques for solving this problem. Section 3 describes the EA development for the PSP problem.

3 Evolutionary Algorithms to Protein Structure Prediction

Evolutionary algorithms (EAs) are population-based metaheuristics which use biology-inspired mechanisms like mutation, crossover, natural selection, and survival of the fittest in order to refine a set of candidate solutions iteratively [2].

The idea behind EAs is the following: given an optimization function, the algorithm creates a set of random candidate solutions (or a *population* of *individuals*). Based on the quality (named *fitness*) of these solutions, some of the best candidates can be chosen to originate the next generation by applying reprodutive operators: *crossover* and/or *mutation*. Crossover is applied to two individuals (so-called *parents*) and results in one or more individuals. Mutation is applied to one individual, resulting in one new individual. Crossover and mutation lead to a set of new individuals (so-called *offspring*) that compete with the parent population — based on the fitness — to create the next generation. The process is iterated until a previously defined condition is reached. Algorithm 1 provides the general scheme of an EA.

Algorithm 1. Pseudocode of a typical EA.

```
   Input: A set of EA parameters (see [2]).
   Output: A population P of the best individuals.
1  begin
2      t ← 0;
       /* initialize a random population of individuals                    */
3      Initialize P(t);
       /* evaluate the individuals in P(t)                                 */
4      Evaluate P(t);
       /* test of termination conditions (for instance, a level of adaptation or a maximum
          value for t)                                                     */
5      while not terminated do
6          t ← t + 1;
           /* find a new population with the best individuals of P(t − 1)  */
7          P'(t) ← Select P(t − 1);
           /* apply crossover to selected individuals                      */
8          P''(t) ← Recombine P'(t);
           /* apply a stochastically change to individuals of the recombined population */
9          P'''(t) ← Mutate P''(t);
           /* evaluate the structures in P'''(t)                           */
10         Evaluate P'''(t);
           /* select the survivors between P(t − 1) and P'''(t)            */
11         P(t) ← Survivors(P(t − 1), P'''(t));
12     end
13 end
```

In the HP model, a protein conformation must be represented in a particular lattice; thus, each individual of the EA represents a conformation. In general, the fold is expressed as a sequence of *moves* into the lattice. The position of the first amino acid is fixed — in cubic lattice, for example, this position is, in general, $(0, 0, 0)$ — and the other positions are specified by $n - 1$ moves for a sequence of n amino acids.

Two major schemes for representing the moves can be found in the literature. In the *absolute representation* (AR) [8] an external reference system is assumed and the moves are specified based on this reference. For example, in cubic lattices, there exist six possibilities of moves: *up* (U), *down* (D), *right* (R), *left* (L), *front* (F), and *back* (B). However, absolute representation allows for the generation of "moves of return", i.e., moves that annuls the previous move (for example, if a move D follows a move U, D will be a move return, because D annul the U move). For this reason, the *relative representation* (RR) [6] was proposed, with a reference system depending on the last move. Thus, if we have a move U, for example, the next move needs to be in the set $\{U, R, L\}$ in square lattices and in $\{U, R, L, F, B\}$, in cubic ones.

On the other hand, relative representation does not avoid unfeasible solutions, since two residues can collide in another position in the lattice. Moreover, standard reproductive operators can generate unfeasible offspring from feasible parents, for example. Thus, unfeasible solutions are allowed, but a *penalty function* assigns lower fitness values to these solutions during the evaluation stage [4,5,10].

One example of the use of penalty function is the following: the fitness function evaluates if we have unfeasible solutions. In the affirmative case, the fitness will be the negative value of the number of collisions; otherwise, the fitness is the sum of `HH` contacts. Thus, feasible solutions have positive fitness and unfeasible solution have negative fitness, and the EA will find a solution that maximizes the number of `HH` contacts.

The evolution of the best solution, using the penalty function described in the example, is illustrated in Fig. 2 for sequence `643d.1` used in the literature [4,5,6,8]. We can observe that the average fitness in the first populations is very low, i.e., all individuals have collision in their structure. Moreover, during the evolutionary process, the unfeasible solutions quickly converge to negative values in opposition to the feasible solutions that have very slow convergence. These observations motivated the development of methodologies which reduce or eliminate the occurrence of conformations with collisions.

In order to solve the problem of unfeasible solutions, we propose two approaches, named *direct representation* (see Section 3.1) and *conformation matrix* (see Section 3.2).

3.1 Direct Representation

Direct representation (DR) is based on relative representation, but it uses three arrays to encode the points in Euclidean space. The first array (called *movx*) encodes x-axis of space by generating three possibilities: -1, representing the left (L) move, i.e., decreasing the x value; 1, which represents the right (R) move, increasing the x value; and 0, which represents no move in x. In this case, a move is generated in the *movy* array, encoding the y-axis: -1 encodes down (D) move, 1 encodes up (U) move and 0 does not encode any move. Finally, the third array (*movz*) represents moves in z-axis, and can assume the values: -1, back (B), 1, front (F) and 0, no move.

Based on these arrays, an individual is created as follows: first, the values of x are randomly generated. If the ith value is different from 0, the ith position in y and z receives 0. Moreover, if this number the in ith position is different from the previous number (position $i-1$ of the array), a new value needs to be

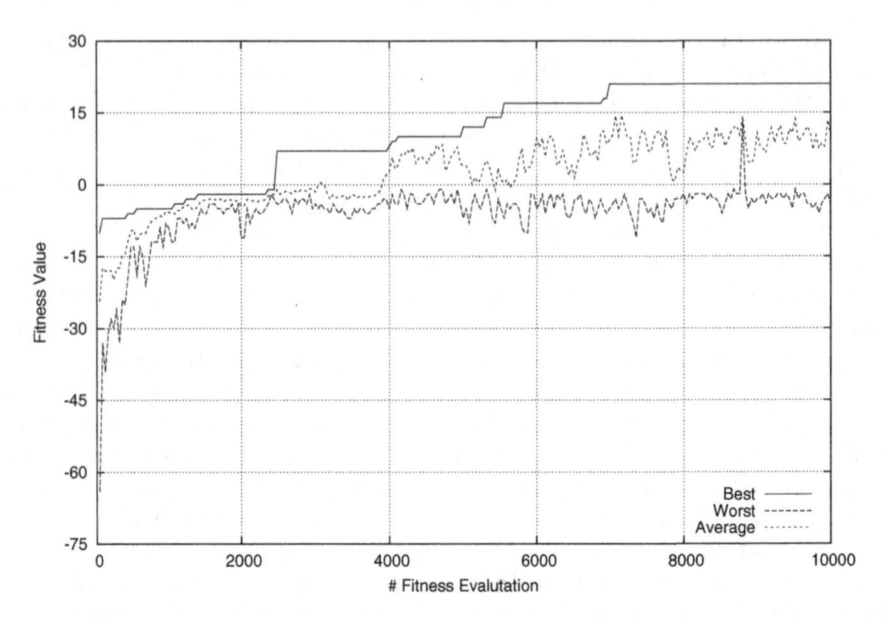

Fig. 2. Number of fitness evaluation by fitness value to the sequence `643d.1`

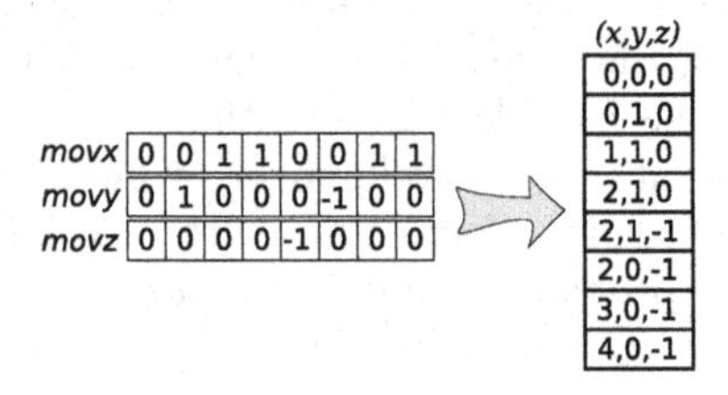

Fig. 3. Example of an individual encoded with direct representation and the spatial positions after the decodification process

generated. This step is important to avoid moves of return in the space; thus, move 1 will never follow move −1, and vice-versa.

If the value in *movx* is 0, there are no moves in x-axis and the algorithm randomly generates a value in *movy*, applying the same methodology stated before. If the value generated in *movy* is 0, a new value will be created in *movz* array. For this reason, this array can receive only two values: 1 and −1. Fig. 3 illustrates the direct representation.

3.2 Conformation Matrix

The second representation proposed uses a matrix (called *conformation matrix*, C_M) to decode the individuals in order to avoid unfeasible solutions and, at the same time, evaluates the number of `HH` contacts. Each position of conformation is indexed in a position of the matrix which represents the lattice. If an indexed

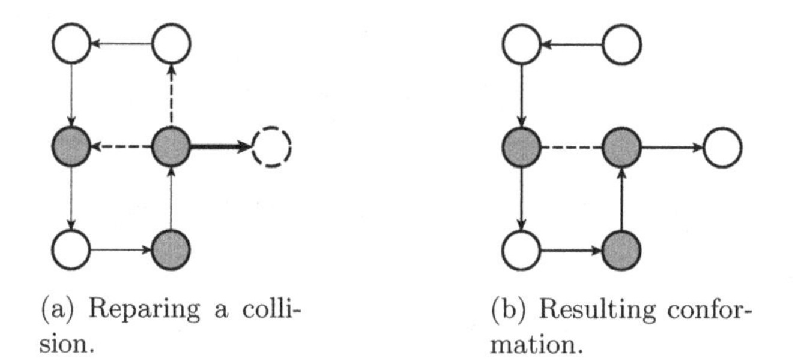

(a) Reparing a collision.

(b) Resulting conformation.

Fig. 4. Illustration of the process of search for feasible solutions: 4(a) shows the potential collision (dashed arrow) and the new position (thick arrow); 4(b) shows the resulting conformation and HH contact (dashed line)

position has an element, i.e., if there is an amino acid in position (x, y, z) of C_M, the algorithm identifies a collision and generates a new value to be decoded in another position of C_M. Fig. 4 illustrates part of a 2-dimensional C_M, where a collision is repaired.

To guarantee that all individuals will be repaired in only one scanning, an efficient procedure must be applied. We use an array of permutations that consists of a set of moves (that are encoded in integers numbers). For each repair after a collision, a new random permutation of this array is created, in order to avoid applying the same sequence of moves. Thus, when an individual is created, a permutation is generated and the first value of the array is used. If this value is invalid, the next value is used. If all possibilities in the array have been explored and a feasible solution has not been found, the individual is eliminated, i.e., the building process is canceled and a new individual starts to be generated.

4 Results and Discussion

This section presents instances illustrating the effect of the proposed approaches to reduce the unfeasible solutions. The initial population of the proposed techniques (direct representation and conformation matrix) was compared with an EA (which starts with a random initial population) based on absolute representation. Moreover, the fitness value was assigned by the procedure described in Section 3, i.e., feasible solutions have positive fitnesses and unfeasible solutions have negative ones.

Table 1 compares the fitness value of the initial population in each case, based on 20 instances: ten sequences with 27 amino acids and ten sequences with 64 amino acids. These sequences were created by [8] and are used as benchmarks in the literature [4,5,6].

Table 1 shows an improvement on fitness average of initial population using both representations. Although, DR allows unfeaseble individuals, they have

Table 1. Comparison of the fitness values of the initial population using the direct representation (DR) and conformation matrix (C_M). The fitness was compared with the absolute representation (AR) and with the best value found by [6].

	Sequence	AR			DR			C_M			Best in [6]
		best	avg.	worth	best	avg.	worth	best	avg.	worth	
	273d.1	1	−7.03	−19	2	−1.15	−7	3	0.37	0	9
	273d.2	2	−7.02	−19	2	−0.95	−7	3	0.62	0	10
	273d.3	1	−7.03	−19	2	−0.93	−10	4	0.93	0	8
	273d.4	1	−7.80	−34	2	−1.00	−10	3	0.56	0	15
	273d.5	−1	−8.77	−22	4	−1.31	−12	3	0.55	0	8
	273d.6	1	−7.03	−19	3	−0.89	−10	5	1.10	0	11
	273d.7	0	−7.04	−19	1	−1.03	−10	2	0.16	0	13
	273d.8	0	−7.81	−34	2	−0.99	−10	3	0.35	0	4
	273d.9	0	−7.04	−19	2	−1.06	−10	3	0.30	0	7
	273d.10	0	−7.81	−34	3	−0.81	−7	3	0.75	0	11
Average		0.50	−7.44	−23.80	2.30	−1.01	−9.30	3.20	0.57	0	
	643d.1	−9	−23.52	−64	2	−4.68	−18	8	1.33	0	27
	643d.2	−9	−23.52	−64	3	−4.02	−22	5	1.32	0	30
	643d.3	−7	−24.69	−67	5	−3.88	−22	4	1.27	0	38
	643d.4	−7	−24.69	−67	4	−4.79	−18	7	1.77	0	34
	643d.5	−7	−23.50	−52	3	−4.76	−18	9	3.47	0	36
	643d.6	−7	−24.69	−67	4	−4.05	−22	8	2.14	0	31
	643d.7	−9	−23.52	−64	4	−3.98	−22	7	2.93	0	25
	643d.8	−7	−23.50	−52	4	−4.35	−28	6	1.86	0	34
	643d.9	−9	−23.52	−64	4	−4.76	−18	8	2.35	0	33
	643d.10	−7	−24.69	−67	3	−4.36	−28	7	1.6	0	26
Average		−7.80	−23.98	−62.80	3.60	−4.36	−21.60	6.90	2.00	0	

higher fitness value than the unfeaseble individuals generated by AR. On the other hand, C_M generates only valid conformations and the individual-repair algorithm provides more adequated population in terms of fitness values. The histograms of Figs. 5 and 6 show the fitness value of the initial population in sequences `273d.1` and `643d.1`, in order to illustrate the behavior of the individuals.

It is possible to observe that the standard EA generates a large number of unfeasible solutions (in Fig. 6(a), for example, all individuals have a negative fitness). The direct representation allows for fewer unfeasible solutions, since return moves have been avoided. Moreover, the average fitness of the population is higher than the random generation. The use of three arrays, however, demands the development of specific reprodutive operators.

On the other hand, the use of conformation matrix builds only feasible individuals. However, several solutions have relatively low fitness because they have few `HH` contacts. To find better individuals (i.e., with more `HH` contacts), we can use the standard evolutionary operators and check the resulting offspring.

An accurate investigation into these new approaches is necessary to evaluate EAs in terms of running time and assessments of objective functions.

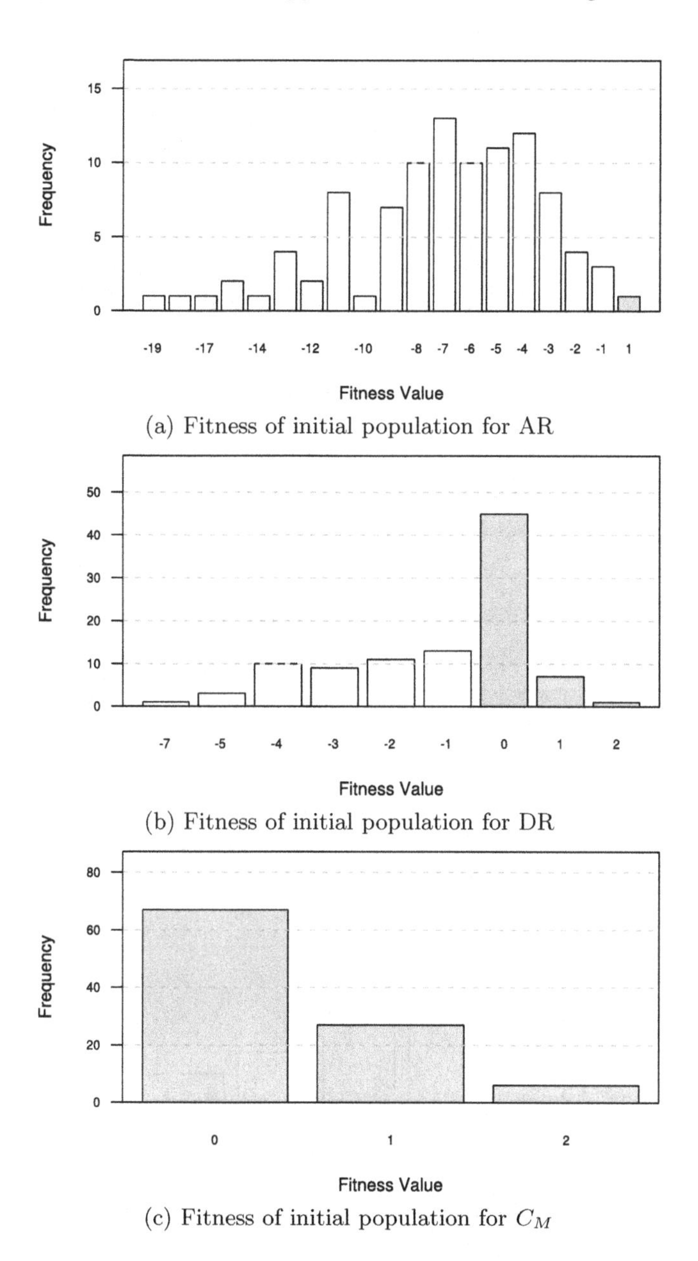

(a) Fitness of initial population for AR

(b) Fitness of initial population for DR

(c) Fitness of initial population for C_M

Fig. 5. Comparison of fitness distribution of individuals in the initial population for the sequence `273d.1`. Dark bars represent positive fitnesses, i.e., feasible solutions.

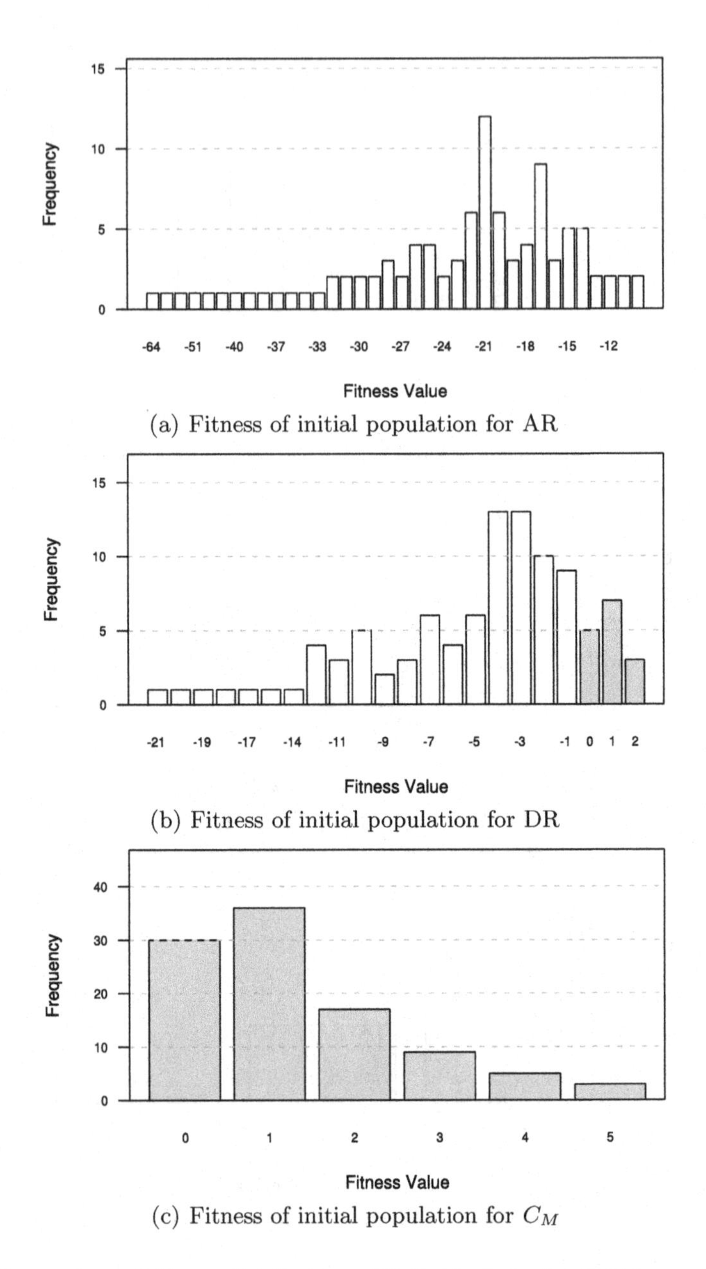

(a) Fitness of initial population for AR

(b) Fitness of initial population for DR

(c) Fitness of initial population for C_M

Fig. 6. Comparison of fitness distribution of individuals in the initial population for the sequence 643.1. Dark bars represent positive fitnesses, i.e., feasible solutions.

5 Conclusions

EAs are usually applied to PSP problem via penalty functions. On the one hand, this approach favors the development of generic algorithms; on the other hand, the unfeasible solutions can decrease the algorithm's performance. This paper has been aimed the study of the effect of two new encodes to create a initial population of EAs with promising regions of search space:

1. Direct representation, which simplifies the encoding process by the use of three-array individual;
2. Conformation matrix, which repairs all unfeasible individuals.

Both strategies where motivated by the fact that EAs are generic search and optimization procedures, therefore they can be biased by an inadequate encoding of the solutions. The representations proposed in this paper are specific for PSP problems and the results based on initial population show that both techniques generated promising candidate solutions. For this reason, both representations will be explored to elaborate more adequate EAs.

Moreover, specific representations of EAs have provided better results in several combinatorial problems [11]. For this reason, the new representations may contribute in the development of more adequated EAs for PSP. Further research on the combination of the proposed representations could also be explored. Another avenue for future research is developing a strategy to employ more complex models of PSP.

Acknowledgments. The authors would like to acknowledge FAPESP (Brazilian research foundation) for the financial support given to this project. They are also indebted to Telma W. L. Soares and Vinicius V. de Melo for their thoughtful discussion and suggestions for improvements.

References

1. Setubal, J.C., Meidanis, J.: Introduction to Computational Molecular Biology. PWS Publishing Company, Boston (1997)
2. De Jong, K.A.: Evolutionary Computation: A Unified Approach. The MIT Press, Cambridge (2006)
3. Dill, K.A.: Theory for the folding and stability of globular proteins. Biochemistry 24(6), 1501–1509 (1985)
4. Khimasia, M.M., Coveney, P.: Protein structure prediction as a hard optimization problem: The genetic algorithm approach. Molecular Simulation 19, 205–226 (1997)
5. Krasnogor, N., Hart, W.E., Smith, J., Pelta, D.A.: Protein structure prediction with evolutionary algorithms. In: Banzhaf, W., Daida, J., Eiben, A.E., Garzon, M.H., Honavar, V., Jakiela, M., Smith, R.E. (eds.) Proceedings of the Genetic and Evolutionary Computation Conference, Orlando, Florida, USA, vol. 2, pp. 1596–1601. Morgan Kaufmann, San Francisco (1999)
6. Patton, A.L., Punch III, W.F., Goodman, E.D.: A standard GA approach to native protein conformation prediction. In: Eshelman, L. (ed.) Proceedings of Sixth International Conference on Genetic Algorithms, pp. 574–581. Morgan Kaufmann, San Francisco (1995)
7. Piccolboni, A., Mauri, G.: Application of evolutionary algorithms to protein folding prediction. In: Hao, J.-K., Lutton, E., Ronald, E., Schoenauer, M., Snyers, D. (eds.) AE 1997. LNCS, vol. 1363, pp. 123–136. Springer, Heidelberg (1998)

8. Unger, R., Moult, J.: A genetic algorithm for 3D protein folding simulations. In: Proceedings of Fifth Annual International Conference on Genetic Algorithms, San Francisco, CA, USA, pp. 581–588 (1993)
9. Clote, P., Backofen, R.: Computational Molecular Biology: An Introduction. Wiley Series in Mathematical and Computational Biology. John Wiley & Sons Inc., New York (2000)
10. Cotta, C.: Protein structure prediction using evolutionary algorithms hybridized with backtracking. In: Mira, J., Álvarez, J.R. (eds.) IWANN 2003. LNCS, vol. 2687, pp. 321–328. Springer, Heidelberg (2003)
11. Rothlauf, F.: Representation for Genetic and Evolutionary Algorithms. Springer, Heidelberg (2006)
12. de Melo, V.V., Delbem, A.C.B., Pinto Júnior, D.L., Federson, F.M.: Discovering promising regions to help global numerical optimization algorithms. In: Gelbukh, A., Kuri Morales, Á.F. (eds.) MICAI 2007. LNCS, vol. 4827, pp. 72–82. Springer, Heidelberg (2007)
13. de Melo, V.V., Delbem, A.C.B.: On promising regions and optimization effectiveness of continuous and deceptive functions. In: IEEE Congress on Evolutionary Computation, June 2008, pp. 4184–4191 (2008)
14. Hart, W.E., Newman, A.: Protein structure prediction with lattice models. In: Aluru, S. (ed.) Handbook of Molecular Biology. Chapman & Hall/CRC Computer and Information Science Series, pp. 1–24. CRC Press, New York (2006)
15. Berger, B., Leighton, T.: Protein folding in the hydrophobic–hydrophilic (HP) model is NP-complete. Journal of Computational Biology 5(1), 27–40 (1998)
16. Crescenzi, P., Goldman, D., Papadimitriou, C., Piccolboni, A., Yannakakis, M.: On the complexity of protein folding. Journal of Computational Biology 5(3), 423–466 (1998)

Comparing Methods for Multilabel Classification of Proteins Using Machine Learning Techniques

Ricardo Cerri, Renato R.O. da Silva, and André C.P.L.F. de Carvalho

Instituto de Ciências Matemáticas e de Computação - ICMC/USP
Avenida Trabalhador São-carlense – 400 – Centro
Caixa Postal 668 – CEP: 13560-970 – São Carlos - SP - Brasil
{cerri,rros,andre}@icmc.usp.br

Abstract. Multilabel classification is an important problem in bioinformatics and Machine Learning. In a conventional classification problem, examples belong to just one among many classes. When an example can simultaneously belong to more than one class, the classification problem is named multilabel classification problem. Protein function classification is a typical example of multilabel classification, since a protein may have more than one function. This paper describes the main characteristics of some multilabel classification methods and applies five methods to protein classification problems. For an experimental comparison of these methods, traditional machine learning techniques are used. The paper also compares different evaluation metrics used in multilabel problems.

Keywords: Machine Learning, Bioinformatics, Multilabel, Classification, Proteins.

1 Introduction

A large number of important data analysis tasks in bioinformatics use classification models. Classification is one of the most investigated issues in machine learning (ML) and data mining research. A classification problem can be formally defined as: given a set of examples consisting of pairs $\{x_i,y_i\}$, find a function that maps each x_i to its associated class y_i, such that $i = 1, 2, ..., m$, where m is the number of training examples.

Most of the classification problems described in the literature refer to single-label problems, where each example is associated with just one among two or more classes. There are classification problems, however, where an example can belong to more than one class simultaneously. These problems are named multilabel classification problems.

Different methods can be used to deal with these problems. Some methods transform a multilabel classification problem into a set of single-label classification problems, using traditional classification algorithms. Other methods develop specific algorithms for the multilabel problem, through modifications in the internal mechanisms of traditional classification algorithms or developing new algorithms. This paper presents a comparison between some of these methods,

K.S. Guimarães, A. Panchenko, T.M. Przytycka (Eds.): BSB 2009, LNBI 5676, pp. 109–120, 2009.

using protein classification datasets related to the Yeast organism and structural families. Different evaluation metrics are also presented and used.

The paper is organized as follows: section 2 presents the main concepts of multilabel classification; section 3 describes the methods used; section 4 shows the evaluation metrics apllied; the materials and methods are reported in section 5; the experiments are shown in section 6, with an analysis of the results obtained. Finally, section 7 presents the main conclusions of this work.

2 Multilabel Classification

In single-label classification problems, a classifier is trained using a set of examples associated with just one class l of a set of classes L, where $|L| > 1$. If $|L| = 2$, the problem is named binary classification problem. If $|L| > 2$, the problem is named multiclass classification problem [1].

In multilabel classification problems, each example can be associated with a set of classes. A multilabel classifier can be defined as a function $H : X \rightarrow 2^L$ that maps an example $x \in X$ in a set of classes $C \in 2^L$.

In the past, the study of multilabel classification models was motivated mainly by text classification and medical diagnostics problems. In a text classification problem, each document can simultaneously belong to more than one topic. An article, for instance, may be simultaneously classified in the topics of Computer Science and Physics. In medical diagnostics, a patient may suffer from diabetes and hypertension at the same time. Multilabel classification works can be found in several application areas, like text classification [2,3,4], bioinformatics [5,6,7,8,9], medical diagnostics [10] and scene classification [11,12]. Figure 1(b) illustrates an example of multilabel classification. The objects belonging simultaneously to classes "●" and "■" are represented by "◇".

Different methods have been proposed in the literature to deal with multilabel classification problems. In one of these methods, single-label classifiers are combined to solve the multilabel problem. Other methods modify single-label classifiers, adapting their internal mechanisms, to allow their use in multilabel problems. Yet, new algorithms can be developed specifically for multilabel problems. The next section describes the methods used in this work to deal with the multilabel classification problems.

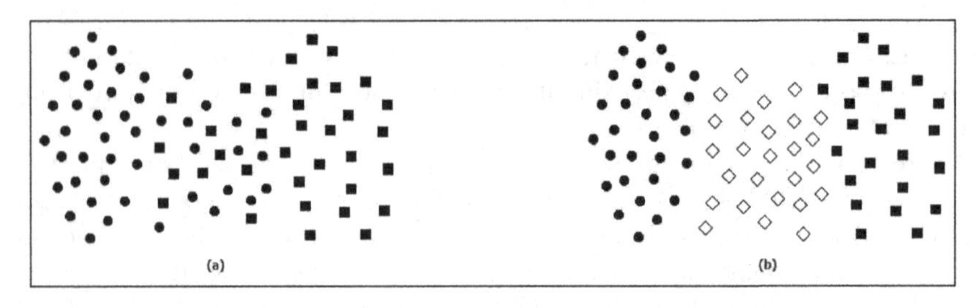

Fig. 1. (a) Tipical classification problem. (b) Multilabel classification problem [12].

3 Multilabel Methods

The methods used in this work follow two different approaches. The first approach is named algorithm independent and, the second, algorithm dependent.

3.1 Algorithm Independent Approach

In this approach, any traditional classification algorithm can be used to deal with the multilabel problem. The idea is to transform the original multilabel problem into a set of single-label classification problems. These transformations can be based on the labels of the examples, named label based transformations, or in the training examples themselves, named example based transformations.

Label Based Transformation. In this transformation, N classifiers are used, where N is the number of classes in the classification problem. Each classifier is associated with one class and is trained to solve a binary classification problem, where its associated class is considered against all other classes. This method is also named Binary Relevance method or One-Against-All [13].

If a multilabel problem with three classes is considered, as each classifier is associated with one class, three classifiers are trained. The problem is therefore divided into three binary classification problems, one for each class. The ith classifier is trained to consider the examples belonging to the ith class as positives and the other examples as negatives. Thus, each classifier becomes specialized in the classification of one particular class. When a new example is presented, the classes for which the classifiers produce a positive output are assigned to it.

A drawback of this method is the fact that it assumes that the classes assigned to an example are independent. This is not always true and, by ignoring the possible correlations between the classes, the result may be poor generalization ability. The process of transformation, though, is reversible, making possible to recover the classes of the original problem from the new created problem.

Example Based Transformation. This transformation redefines the set of classes associated with each example, converting the original multilabel problem into one or more single-label problems. Unlike the previous transformation, this transformation does not produce only binary problems. It may produce both binary or multiclass problems. In this work, three different methods based on this tranformation were used: label creation, random k-labelsets and label conversion.

- **Label Creation.** In this method, for each example, all the classes assigned to this example are combined to form a new and unique class. With this combination, the number of classes involved in the problem may increase considerably, and some classes may have few examples. Supposing a dataset with multilabel examples belonging to classes A and B, these labels would be combined, for example, in a new label C_{AB}.

  This method was named Label-Powerset in [13], and the same name is being adopted in this paper. The creation of new classes allows the recovery of the classes in the original problem.

- **RAndom k-LabELsets.** Based on the Label-Powerset method, a method named RAKEL (RAndom k-LabELsets) [13] is also used in this work. This method iteratively builds a combination of m Label-Powerset classifiers. Given $L = \{\lambda_i\}, i = 1..|L|$ the set of labels of the problem, a *k-labelset* is given by a subset $Y \subseteq L$, with $k = |Y|$. The term L^k represents the set of all *k-labelsets* of L. At each iteration, $1..m$, a *k-labelset* Y_i is randomly selected from L^k without replacement. A classifier H_i is then trained for Y_i.

  For the classification of a new example x, each classifier H_i makes a binary decision for each label λ_j of the corresponding *k-labelset* Y_i. An average decision is calculated for each label λ_j in L, and the final decision is positive for a given label if the average decision is larger than a given threshold t.

  The proposal of RAKEL is to take into account the correlations between classes and, at the same time, avoid the drawback of the Label-Powerset method, where some classes end up with few examples.

- **Label Conversion.** This method uses a label decomposition process, where all multilabel examples are decomposed into a set of single-label examples. In this process, for each example, each possible class is considered as the positive class in sequence, using the multilabel data more than once during the training phase. If the dataset has multilabel examples with labels A, B and C, when a classifier for class A is trained, all the multilabel examples that belong to class A become single-label examples for class A, and the same happens for the other classes. The method was proposed by [12] and is named Cross-Training.

  The number of classifiers is equal the number of classes that belong to at least one multilabel example. This method permits the recovery of the original multilabel data.

3.2 Algorithm Dependent Approach

In this approach, new algorithms are proposed specifically for multilabel classification. It is claimed that a specific algorithm, developed for multilabel classification problems, may have a better performance than methods based on the algorithm independent approach. In this paper, a method proposed by [6], named ML-kNN, is used. In this method, for each example, the classes associated with the k nearest neighbor examples are recovered, and a counting of the neighbors associated to each class is made. According with the set of labels associated with the neighbor examples, the *maximum a posteriori* principle [14] is used to define the set of classes for a new example.

The methods investigated in this work need to be evaluated regarding their predictive accuracy. The next section presents the metrics adopted in this work for the evaluation and comparison of the classification methods.

4 Evaluation Metrics

The evaluation of multilabel classifiers requires metrics different from those used in single-label problems. Whereas in a single-label problem the classification of an example is correct or incorrect, in a multilabel problem a classification of an example may be partially correct or partially incorrect. This situation could happen when a classifier correctly assigns to an example at least one of the classes it belongs to, but does not assign one or more classes it also belongs to. A classifier could also assign to an example one or more classes it does not belong to.

The criterion of evaluation used in this paper is based on the multilabel classification provided by the classifier being evaluated. The evaluation uses the labels assigned to a given example by the classifier, comparing them with the true labels of the example.

Given a set of multilabel data D, consisting of $|D|$ multilabel examples (x_i, Y_i), with $i = 1...|D|$ and $Y_i \subseteq L$, and H a multilabel classifier with $Z_i = H(x_i)$ being the set of predicted classes for a given example x_i. A very commom metric, used in the work of [15], is the Hamming Loss, which is defined in Equation 1.

$$HammingLoss(H, D) = \frac{1}{|D|} \sum_{i=1}^{|D|} \frac{|Y_i \Delta Z_i|}{|L|} \tag{1}$$

In Equation 1, Δ represents the symmetric difference between two sets, and corresponds to the XOR operation from boolean logic [1]. The lower the value of Hamming Loss, the better is the classification. Its value varies from zero to one, and a perfect situation occurs when it is equal to zero.

Other metrics used were Accuracy, Precision and Recall, employed in the work of [16]. They are presented in Equations 2, 3 and 4. The Accuracy symmetrically measures how close Y_i is to Z_i. Precision can be defined as the percentage of true positive examples from all the examples classified as positive by the classification model, and Recall is the percentage of examples classified as positive by a classification model that are true positive.

$$Accuracy(H, D) = \frac{1}{|D|} \sum_{i=1}^{|D|} \frac{|Y_i \cap Z_i|}{|Y_i \cup Z_i|} \tag{2}$$

$$Precision(H, D) = \frac{1}{|D|} \sum_{i=1}^{|D|} \frac{|Y_i \cap Z_i|}{|Z_i|} \tag{3}$$

$$Recall(H, D) = \frac{1}{|D|} \sum_{i=1}^{|D|} \frac{|Y_i \cap Z_i|}{|Y_i|} \tag{4}$$

Two other metrics adopted in this work, named F-Measure and Subset Accuracy, were used in [13]. F-Measure is a combination of Precision and Recall. This combination is the harmonic average of the two metrics and is used as an aggregated performance score. Subset Accuracy is a very restritive accuracy metric, considering a classification as correct only if all the labels predicted by a classifier are correct. These metrics are illustrated in Equations 5 and 6.

$$F - Measure(H, D) = \frac{1}{|D|} \sum_{i=1}^{|D|} \frac{2|Y_i \cap Z_i|}{|Z_i| + |Y_i|} \tag{5}$$

$$Sub - Accuracy(H, D) = \frac{1}{|D|} \sum_{i=1}^{|D|} I(|Z_i| = |Y_i|) \tag{6}$$

The values of these metrics vary from zero to one. All the results, except in the Hamming Loss metric, were multiplied by 100 for better visualization.

5 Materials and Methods

5.1 Datasets

The datasets used in the experiments are composed of proteins related to the organism Yeast and protein sequences classified in structural families. Table 1 presents the main characteristics of these datasets.

Yeast is a unicellular fungus and its better known specie is named *Saccharomyces cerevisiae*. This organism is often used in the fermentation of sugar for the production of ethanol, and also in the fermentation of wheat and barley for the production of alcoholic beverages. It has also been employed as vitamin supplement, due to its composition rich in proteins and for being source of vitamins. The dataset used has 2417 examples, 103 attributes and 14 classes.

The dataset of protein structural families has 662 examples, 1186 attributes and 14 classes. The classes are part of important protein families in Biology, like oxydoreductases (enzymes that catalyze the oxidation-reduction reactions), isomerases (enzymes that catalyze the structural rearrangement of isomers) and hydrolases (enzymes that catalyze the hydrolysis of a chemical bond).

These datasets are freely available at `http://mlkd.csd.auth.gr/multilabel.html#Datasets`.

5.2 Implemented Methods

The methods One-Against-All, Label-Powerset and Cross-Training were implemented using the R tool [17]. For each of these methods, the classification algorithms KNN [18], SVM [19], C4.5 [20], Ripper [21] and BayesNet [22] were used.

Table 1. Characteristics of the Datasets

		N. Attributes		N. Examples per Class		
	N. Examples	Numeric	Nominal	Least	Average	Maximum
Yeasts	2417	103	0	34	731.5	1186
Sequences	662	0	1186	17	54.5	171

	N. Multilabel Examples	Class Number Average per Example
Yeasts	2385	4.23
Sequences	69	1.15

The parameter values used in the algorithms were the default values suggested in the R tool. Five neighbors were used in the KNN algorithm.

In addition to the implemented methods, the methods ML-kNN and RAKEL, implemented in the software Mulan [23], were used. The experiments were performed using the k-fold cross-validation methodology with k equal to 5.

5.3 Statistical Tests

In order to evaluate the statistical significance of the results obtained, the corrected Student's t test for paired data is used, which consists in verify if the average of the difference between the results of two classifiers, considering many test sets, is different from zero [24]. The level of significance of the test was determined as 5% ($\alpha = 0.05$). After the Bonferroni correction was applied, the α used in the comparisons had value equal to 0.005. This correction was needed because many comparisons were performed. The Bonferroni correction divides the value of α by the number of tests performed, making the value of α more stringent and decreasing the chance of errors [25]. The tests were performed using four degrees of freedom.

6 Experiments

The experiments were performed to compare the performance of the methods One-Against-All, Label-Powerset, Cross-Training, ML-kNN and RAKEL. The algorithms KNN, SVM, C4.5, Ripper and BayesNet were used in the comparisons, and these algorithms were also compared with each other considering the methods separately. Tables 2, 3, 4, 5 and 6 present the mean and standard deviation of the experimental results obtained using the five methods for the Yeast dataset. Tables 7, 8, 9, 10 and 11 present the results for the protein family dataset. The results were evaluated using six metrics: Hamming Loss, Accuracy, Subset-Accuracy, Precision, Recall and F-Measure. The best results for each metric are shown in boldface.

In the Yeast dataset, considering tables 2, 3, 4 and 5, the best performances were achived, in the majority of the algorithms, by the methods Cross-Training and RAKEL. This occurred because, in the Cross-Training method, the data are used more than once during the training phase, making the data less sparse. The Label-Powerset method achieved the worst results in the majority of the

Table 2. Results Using One-Against-All in the Yeast Dataset

	HammingLoss	Precision	Recall	F-Measure	Accuracy	Subset-Accuracy
KNN	**0.20 (0.006)**	67.14 (1.0)	60.38 (1.5)	60.86 (1.1)	51.02 (1.1)	**20.81 (1.3)**
SVM	**0.20 (0.003)**	67.85 (1.0)	60.49 (1.3)	**61.15 (1.0)**	**51.05 (0.9)**	18.37 (1.2)
C4.5	0.24 (0.005)	60.28 (1.7)	56.07 (2.4)	55.06 (1.7)	42.79 (1.4)	6.40 (1.0)
Ripper	0.21 (0.006)	**69.11 (2.6)**	54.71 (1.0)	57.97 (1.2)	46.32 (1.0)	9.14 (0.5)
BayesNet	0.25 (0.012)	59.06 (2.2)	**60.65 (1.7)**	56.74 (2.0)	45.01 (2.1)	10.26 (2.1)

Table 3. Results Using Label-Powerset in the Yeast Dataset

	HammingLoss	Precision	Recall	F-Measure	Accuracy	Subset-Accuracy
KNN	0.24 (0.005)	60.51 (1.1)	54.83 (1.5)	55.39 (1.0)	46.58 (1.1)	22.21 (2.3)
SVM	**0.20 (0.003)**	**68.03 (1.0)**	**63.50 (1.3)**	**63.48 (0.7)**	**54.55 (0.6)**	**27.88 (1.6)**
C4.5	0.27 (0.006)	53.89 (1.6)	51.33 (1.4)	50.01 (1.1)	40.06 (0.8)	13.15 (1.4)
Ripper	0.25 (0.007)	58.55 (1.0)	54.83 (1.4)	55.27 (1.2)	44.71 (1.5)	15.80 (1.6)
BayesNet	0.27 (0.009)	54.27 (1.4)	50.49 (2.0)	51.41 (1.6)	39.78 (1.5)	8.77 (1.1)

Table 4. Results Using Cross-Training in the Yeast Dataset

	HammingLoss	Precision	Recall	F-Measure	Accuracy	Subset-Accuracy
KNN	**0.22 (000.7)**	63.14 (0.8)	70.96 (1.4)	64.28 (0.9)	53.61 (0.9)	**19.90 (1.4)**
SVM	**0.22 (0.008)**	**63.29 (1.3)**	78.22 (1.7)	**67.12 (1.3)**	**56.26 (1.2)**	19.11 (1.3)
C4.5	0.35 (0.008)	47.24 (1.0)	77.94 (1.4)	56.55 (1.0)	42.56 (0.9)	3.76 (0.6)
Ripper	0.30 (0.005)	51.32 (0.4)	**81.74 (0.4)**	60.78 (0.2)	47.19 (0.3)	4.67 (0.8)
BayesNet	0.25 (0.003)	58.50 (0.7)	69.25 (1.2)	60.90 (0.7)	48.88 (0.7)	11.13 (0.9)

Table 5. Results Using RAKEL in the Yeast Dataset

	HammingLoss	Precision	Recall	F-Measure	Accuracy	Subset-Accuracy
KNN	0.24 (0.008)	59.56 (1.5)	59.48 (1.2)	59.52 (1.3)	47.49 (1.6)	20.35 (2.0)
SVM	0.19 (0.006)	70.38 (1.2)	**64.44 (1.9)**	**67.27 (1.5)**	**55.14 (1.4)**	**24.03 (1.5)**
C4.5	**0.18 (0.003)**	**73.33 (0.8)**	57.61 (0.6)	64.52 (0.6)	51.24 (0.4)	18.45 (1.7)
Ripper	0.21 (0.005)	68.35 (0.5)	54.71 (2.0)	60.76 (1.3)	48.35 (1.3)	16.92 (1.7)
BayesNet	0.22 (0.004)	64.62 (1.2)	59.94 (0.8)	62.21 (0.9)	48.84 (0.8)	15.76 (1.4)

Table 6. Results Using ML-kNN in the Yeast Dataset

HammingLoss	Precision	Recall	F-Measure	Accuracy	Subset-Accuracy
0.19 (0.006)	72.04 (1.3)	58.43 (1.5)	64.51 (1.1)	51.06 (1.0)	17.91 (1.2)

Table 7. Results Using One-Against-All in the Protein Family Dataset

	HammingLoss	Precision	Recall	F-Measure	Accuracy	Subset-Accuracy
KNN	0.002 (0.001)	98.13 (1.2)	97.78 (1.5)	97.90 (1.3)	97.73 (1.4)	96.97 (1.8)
SVM	**0.0005 (0.0003)**	**99.67 (0.3)**	**99.77 (0.3)**	**99.68 (0.2)**	**99.59 (0.2)**	**99.24 (0.5)**
C4.5	0.001 (0.001)	99.11 (0.8)	99.04 (1.1)	98.93 (0.9)	98.61 (1.3)	97.42 (2.7)
Ripper	0.001 (0.0003)	99.11 (0.3)	99.47 (0.5)	99.21 (0.3)	99.04 (0.3)	98.48 (0.5)
BayesNet	0.004 (0.0009)	96.39 (1.0)	98.33 (0.3)	97.00 (0.7)	96.24 (0.8)	93.80 (1.4)

algorithms due to the large number of labels created for the classifiers training (198), and some of these labels have few examples. Even with this disadvantage, this method achieved some of the best results for the Subset-Accuracy metric, because it considers correlations between classes. The RAKEL method had a

Table 8. Results Using Label-Powerset in the Protein Family Dataset

	HammingLoss	Precision	Recall	F-Measure	Accuracy	Subset-Accuracy
KNN	0.003 (0.002)	98.43 (1.2)	98.08 (1.6)	98.20 (1.4)	98.03 (1.5)	97.27 (2.0)
SVM	**0.0006 (0.0008)**	99.79 (0.3)	**99.67 (0.5)**	99.70 (0.4)	**99.62 (0.5)**	**99.24 (0.9)**
C4.5	0.001 (0.002)	99.44 (0.7)	99.41 (0.7)	99.41 (0.7)	99.32 (0.7)	99.09 (0.8)
Ripper	0.003 (0.003)	98.43 (1.3)	98.41 (1.2)	98.40 (1.2)	98.36 (1.2)	98.18 (1.2)
BayesNet	0.0007 (0.0008)	**99.95 (0.1)**	**99.67 (0.4)**	**99.76 (0.2)**	**99.62 (0.4)**	98.94 (1.1)

Table 9. Results Using Cross-Training in the Protein Family Dataset

	HammingLoss	Precision	Recall	F-Measure	Accuracy	Subset-Accuracy
KNN	**0.061 (0.0075)**	77.55 (2.7)	82.51 (2.6)	78.57 (2.6)	74.64 (3.0)	**63.42 (4.3)**
SVM	**0.061 (0.0073)**	77.86 (2.9)	83.57 (1.7)	79.25 (2.3)	75.11 (2.8)	63.27 (4.3)
C4.5	**0.061 (0.0073)**	77.83 (2.9)	83.52 (1.8)	79.23 (2.4)	75.11 (2.8)	63.27 (4.3)
Ripper	0.062 (0.0075)	77.25 (2.9)	82.82 (2.0)	78.60 (2.5)	74.51 (2.9)	62.82 (4.3)
BayesNet	**0.061 (0.0072)**	**78.16 (2.8)**	**83.72 (1.9)**	**79.43 (2.4)**	**75.29 (2.8)**	**63.42 (4.3)**

Table 10. Results Using RAKEL in the Protein Family Dataset

	HammingLoss	Precision	Recall	F-Measure	Accuracy	Subset-Accuracy
KNN	0.002 (0.001)	98.59 (0.9)	98.28 (0.6)	98.43 (0.7)	98.23 (0.6)	97.73 (0.7)
SVM	**0.0006 (0.0004)**	**99.57 (0.4)**	**99.77 (0.3)**	**99.67 (0.3)**	**99.49 (0.3)**	**99.09 (0.6)**
C4.5	0.001 (0.0003)	99.11 (0.3)	99.47 (0.5)	99.29 (0.4)	99.04 (0.4)	98.48 (0.5)
Ripper	0.001 (0.0003)	99.11 (0.3)	99.47 (0.5)	99.29 (0.4)	99.04 (0.4)	98.48 (0.5)
BayesNet	0.003 (0.001)	97.38 (1.1)	98.54 (0.9)	97.95 (0.8)	97.27 (1.0)	95.76 (1.7)

Table 11. Results Using ML-kNN in the Protein Family Dataset

HammingLoss	Precision	Recall	F-Measure	Accuracy	Subset-Accuracy
0.002 (0.001)	97.81 (1.1)	97.86 (1.3)	97.99 (1.2)	97.81 (1.3)	97.28 (1.6)

better performance than the Label-Powerset method due the way it combines many Label-Powerset classifiers, overcoming the problem of sparse classes.

Some of the results obtained by the ML-kNN method (Table 6), were better than those achieved by the other four methods. Compared with the One-Against-All method, ML-kNN performed better in the majority of the algorithms, but no statistical difference was detected between ML-kNN and the algorithms SVM and KNN, with exception of the Precision metric. In the comparison with the method Label-Powerset, ML-kNN performed better, but there was no statistical difference between ML-kNN and SVM, except for the metrics Accuracy and Subset-Accuracy. Experiments also show that SVM obtained better results for these two metrics. The Cross-Training method using the SVM and KNN algorithms performed better than ML-kNN, except in the metrics Precision and Hamming Loss. The SVM algorithm used in method RAKEL performed better than ML-kNN in most of the metrics, with exception of Precision and

Hamming Loss. The C4.5 algorithm also performed better than ML-kNN in these two metrics, but no statistical significance was detected.

Comparing the algorithms considering the methods separately, SVM achieved the best predictive accuracy results in the One-Against-All and Label-Powerset methods. There was no statistical significance in the comparison with KNN in One-Against-All, but in Label-Powerset, however, SVM performance was statistically superior. In the Cross-Training method, SVM achieved the best results, except for the Recall metric, where Ripper performed better, and for the Subset-Accuracy metric, where KNN obtained a better result. In some other metrics, the better performance of SVM was also not statistically significant. In the RAKEL method, SVM also presented the best results, except for the metrics Hamming Loss and Precison, where C4.5 was better, but there was no statistical significance in the Hamming Loss metric. The best pesformance of SVM was statistically significant.

In the protein family dataset, considering tables 7, 8, 9 and 10, the best results were achieved, in the majority of the algorithms, by the Label-Powerset and RAKEL methods. A possible reason is that, in this dataset, unlike the Yeast dataset, there are few multilabel combinations, and the number of labels formed for training was much lower, equal to 15. The high performance obtained by the methods Label-Powerset, One-Against-All and RAKEL may be due the class distribution in the training set. This dataset has less classes per example and less multilabel examples than the Yeast dataset. Despite the difference in the results of the three methods, there was statistical significance only in the difference of the results for the BayesNet algorithm. The worst performance of the Cross-Training method may be due the large number of examples that are not multilabel, which affects the performance of the method, using more unbalanced data in the training phase.

When comparing the ML-kNN method (Table 11) with the four previous methods, the difference in accuracy performances regarding the methods ML-kNN, One-Against-All, Label-Powerset and RAKEL was not statistically significant, with exception of the algorithm BayesNet in the method One-Against-All for the metrics Hamming Loss and Subset-Accuracy. There was statistical significance in all results when comparing ML-kNN with the Cross-Training method.

Considering the methods separately, SVM achieved slightly better results than the other ML algorithms, with exception of Cross-Training method. However, there was practically no statistical difference in the predictive accuracy results. Among the algorithms in the One-Against-All method, there was statistical significance only in the difference of the results between BayesNet and the other ML algorithms. In the Label-Powerset method, there was no statistical significance in the difference of the accuracy results. In the Cross-Training method, there was statistical significance only in the difference of the results between Ripper and BayesNet for the metrics Accuracy, Recall and F-Measure. In the RAKEL method, the difference of the results was statistically significant between the algorithm KNN and the algorithms SVM, C4.5 and Ripper for the Recall metric. In the other metrics, there was statistical significance only between the results of the SVM and BayesNet algorithms.

7 Conclusion

This paper presented a comparison between different methods of multilabel classification using bioinformatics problems. The methods compared follow two approaches: algorithm independent and algorithm dependent. Different evaluation metrics were used in the comparisons. In the experimental results, SVM usually presented the best results on both datasets. This may have ocurred because SVM dealt better with features from the datasets and methods used, which produced scarce training data.

Regarding the multilabel approaches investigated, the performance of the methods was influenced by the datasets distribution, particularly the distribution of classes per example and the number of multilabel examples. This features should thus be considered when choosing the multilabel method.

The use of the algorithm dependent approach can lead to better results, because the absence of the disadvantages found in the algorithm independent approach. The model produced, however, becomes more complex, once it has to deal with all the classes simultaneously. Beyond that, experiments showed that the algorithm independent approach can achieve better or similar results, while allowing the use of many classification algorithms.

As future work, the authors want to add more multilabel methods in the comparisons and adapt some multilabel classification methods to deal with multilabel problems where the classes are hierarchically structured.

References

1. Tsoumakas, G., Katakis, I.: Multi label classification: An overview. International Journal of Data Warehousing and Mining 3(3), 1–13 (2007)
2. Gonçalves, T., Quaresma, P.: A preliminary approach to the multilabel classification problem of portuguese juridical documents. In: Pires, F.M., Abreu, S.P. (eds.) EPIA 2003. LNCS, vol. 2902, pp. 435–444. Springer, Heidelberg (2003)
3. Lauser, B., Hotho, A.: Automatic multi-label subject indexing in a multilingual environment. In: Koch, T., Sølvberg, I.T. (eds.) ECDL 2003. LNCS, vol. 2769, pp. 140–151. Springer, Heidelberg (2003)
4. Luo, X., Zincir-Heywood, N.A.: Evaluation of two systems on multi-class multi-label document classification. In: International Syposium on Methodologies for Intelligent Systems, pp. 161–169 (2005)
5. Clare, A., King, R.D.: Knowledge discovery in multi-label phenotype data. In: Siebes, A., De Raedt, L. (eds.) PKDD 2001. LNCS (LNAI), vol. 2168, pp. 42–53. Springer, Heidelberg (2001)
6. Zhang, M.L., Zhou, Z.H.: A k-nearest neighbor based algorithm for multi-label classification. In: IEEE International Conference on Granular Computing, vol. 2, pp. 718–721. The IEEE Computational Intelligence Society (2005)
7. Elisseeff, A.E., Weston, J.: A kernel method for multi-labelled classification. In: Advances in Neural Information Processing Systems, vol. 14, pp. 681–687. MIT Press, Cambridge (2001)
8. Alves, R., Delgado, M., Freitas, A.: Multi-label hierarchical classification of protein functions with artificial immune systems. In: Advances in Bioinformatics and Computational Biology, pp. 1–12 (2008)

9. Diplaris, S., Tsoumakas, G., Mitkas, P., Vlahavas, I.: Protein classification with multiple algorithms. In: Bozanis, P., Houstis, E.N. (eds.) PCI 2005. LNCS, vol. 3746, pp. 448–456. Springer, Heidelberg (2005)
10. Karalic, A., Pirnat, V.: Significance level based multiple tree classification. Informatica 5 (1991)
11. Boutell, M.R., Luo, J., Shen, X., Brown, C.M.: Learning multi-label scene classification. Pattern Recognition 37(9), 1757–1771 (2004)
12. Shen, X., Boutell, M., Luo, J., Brown, C.: Multi-label machine learning and its application to semantic scene classification. In: International Symposium on Electronic Imaging, San Jose, CA, January 2004, pp. 18–22 (2004)
13. Tsoumakas, G., Vlahavas, I.: Random k-labelsets: An ensemble method for multi-label classification. In: Kok, J.N., Koronacki, J., Lopez de Mantaras, R., Matwin, S., Mladenič, D., Skowron, A. (eds.) ECML 2007. LNCS, vol. 4701, pp. 406–417. Springer, Heidelberg (2007)
14. Saridis, G.: Parameter estimation: Principles and problems. Automatic Control, IEEE Transactions on 28(5), 634–635 (1983)
15. Schapire, R.E., Singer, Y.: Boostexter: a boosting-based system for text categorization. In: Machine Learning, pp. 135–168 (2000)
16. Godbole, S., Sarawagi, S.: Discriminative methods for multi-labeled classification. In: Advances in Knowledge Discovery and Data Mining, pp. 22–30 (2004)
17. R Development Core Team: R: A Language and Environment for Statistical Computing. R Foundation for Statistical Computing, Vienna, Austria (2008) ISBN 3-900051-07-0
18. Aha, D.W., Kibler, D., Albert, M.K.: Instance-based learning algorithms. Machine Learning 6(1), 37–66 (1991)
19. Vapnik, V.N.: The Nature of Statistical Learning Theory (Information Science and Statistics). Springer, Heidelberg (1999)
20. Quinlan, J.R.: C4.5: programs for machine learning. Morgan Kaufmann Publishers Inc., San Francisco (1993)
21. Cohen, W.W.: Fast effective rule induction. In. Proceedings of the Twelfth International Conference on Machine Learning, pp. 115–123 (1995)
22. Friedman, N., Geiger, D., Goldszmidt, M.: Bayesian network classifiers. Mach. Learn. 29(2-3), 131–163 (1997)
23. Tsoumakas, G., Friberg, R., Spyromitros-Xioufis, E., Katakis, I., Vilcek, J.: Mulan software - java classes for multi-label classification (May 2008), `http://mlkd.csd.auth.gr/multilabel.html#Software`
24. Demšar, J.: Statistical comparisons of classifiers over multiple data sets. Journal of Machine Learning Research 7, 1–30 (2006)
25. Abdi, H.: Bonferroni and Sidak corrections for multiple comparisons. Encyclopedia of Measurement and Statistics, pp. 175–208. Sage, Thousand Oaks (2007)

Comparative Study of Classification Algorithms Using Molecular Descriptors in Toxicological DataBases

Max Pereira[1], Vítor Santos Costa[3], Rui Camacho[1], Nuno A. Fonseca[2,3], Carlos Simões[4], and Rui M.M. Brito[4]

[1] LIAAD-INESC Porto LA & FEUP, Universidade do Porto, Rua Dr Roberto Frias s/n, 4200-465 Porto, Portugal
[2] Instituto de Biologia Molecular e Celular (IBMC), Universidade do Porto Rua do Campo Alegre 823, 4150-180 Porto, Portugal
[3] CRACS-INESC Porto LA, Universidade do Porto, Rua do Campo Alegre 1021/1055, 4169-007 Porto, Portugal
[4] Chemistry Department, Faculty of Science and Technology and Center for Neuroscience and Cell Biology University of Coimbra, Portugal

Abstract. The rational development of new drugs is a complex and expensive process, comprising several steps. Typically, it starts by screening databases of small organic molecules for chemical structures with potential of binding to a target receptor and prioritizing the most promising ones. Only a few of these will be selected for biological evaluation and further refinement through chemical synthesis. Despite the accumulated knowledge by pharmaceutical companies that continually improve the process of finding new drugs, a myriad of factors affect the activity of putative candidate molecules *in vivo* and the propensity for causing adverse and toxic effects is recognized as the major hurdle behind the current "target-rich, lead-poor" scenario. In this study we evaluate the use of several Machine Learning algorithms to find useful rules to the elucidation and prediction of toxicity using 1D and 2D molecular descriptors. The results indicate that: i) Machine Learning algorithms can effectively use 1D molecular descriptors to construct accurate and simple models; ii) extending the set of descriptors to include 2D descriptors improve the accuracy of the models.

1 Introduction

The amount of information concerning chemicals that is available in databases has been increasing at a considerable pace in the last years, changing the whole process of discovery and development of new drugs. These databases have been used as a starting point for screening candidate molecules, and enable the pharmaceutical industry to produce over 100,000 new compounds per year [1]. The promising compounds are further analysed in the development process, where, among other investigations, their potential toxicity is assessed. This is a complex and costly process that often requires years before compounds can be tested in

K.S. Guimarães, A. Panchenko, T.M. Przytycka (Eds.): BSB 2009, LNBI 5676, pp. 121–132, 2009.

human subjects [2,3,4]. Additionally, about 90% of the initially considered drugs fail to reach the market due to toxicological properties [5]. This fact highlights the importance of determining as early as possible toxicological features.

Toxicity tests determine whether or not a candidate molecule is likely to produce toxic effects in humans, usually involve the use of animal models at a pre-clinical stage. As the number of of biological targets identified as relevant increases (resulting from the humane genome project), and hence the demand for drug screening campaigns there is a growing need of efficient *in silico* methods for the prediction of toxicity of organic compounds. The problem is to identify clear relationships between a molecule's chemical structure and its toxicological activity. These relationships can be used to build predictive models to apply to new compounds [6]. Ultimately, this task can be regarded as a method to predict Quantitative Structure-Activity Relationships (QSARs) [7] which considers a set of structural features associated with a toxicity endpoint. These models offer an inexpensive and fast way of estimating molecules toxicological properties [8,9].

The problem of estimating the toxicity of drugs has been addressed, mainly, from three methods: i) regression from physical-chemical properties; ii) expert systems and; iii) machine learning [10,11]. Some toxicity prediction programs are commercially available including TOPKAT (toxicity-prediction by computer-assisted technology), DEREK (deductive estimation of risk from existing knowledge), CSGenoTox, MetaDrug and HazardExpert [12]. These programs have a common characteristic, they are classified as "global" models [8] since they were developed using a non-congeneric set of chemicals. Actually it is not mandatory that the chemicals in these data sets are congeneric, but they should share structural features. Besides the commercially available programs, other studies have been published using machine learning approaches [3,13,10,6,11].

In this paper we compare the performance of classification algorithms in predicting the toxicity of compounds and present the results for data sets composed of constitutional molecular descriptors exclusively (*1D descriptors*) and data sets composed of both constitutional and topological molecular descriptors (*2D descriptors*). Although similar studies have been reported [3,4,10,14], they did not assess the relevancy of molecular descriptors in terms of toxicity prediction. We have applied the classification algorithms to predict the toxicity of compounds for the estrogen receptor (DSSTox NCTRER DataBase), mutagenicity using the carcinogenic potency database (DSSTox CPDBAS DataBase), the Fathead Minow Acute Toxicity (DSSTox EPAFHM DataBase) and the Disinfection By-products Carcinogenicity Estimates (DSSTox DBPCAN DataBase).

In this study we aim at proving/disproving the following hypotheses:

HC0: 1D descriptors contain sufficient information for Machine Learning algorithms to construct accurate and robust predictive models that can predict *whether* a given molecule is toxic.

HC1: Extending the set of 1D descriptors with 2D descriptors improves the accuracy of the models produced by Machine Learning algorithms to predict the degree of toxic activity of molecules.

HR0: 1D descriptors contain sufficient information for Machine Learning algorithms to construct accurate and simple predictive models of the *degree* of toxic activity of molecules.

HR1: Extending the 1D set of descriptors of molecules with 2D descriptors improves the models constructed by Machine Learning algorithms to predict the *degree* of toxic activity of molecules.

The remaining portion of the paper has the following structure. Section 2.1 describes the data sets used, and Section 2.2 gives an overview on the Machine Learning algorithms used in this study. Section 3 details the experiments undertaken and discussed the results obtained. We compare our work with previously published work in Section 4. Conclusions are presented in the last section of the paper.

2 Material and Methods

2.1 Data Sets

We used four data sets available from the Distributed Structure-Searchable Toxicity (DSSTox) Public DataBase Network [15] from the U.S.Environmental Protection Agency[1]. The DSSTox database project is targeted to toxicology studies and uses a standard chemical structure annotation. The data sets are characterised as follows.

CPDB: The Carcinogenic Potency DataBase (CPDB) contains detailed results and analyses of 6540 chronic, long term carcinogenesis bioassays [16], which currently contains 1547 compounds. For the purpose of this study the carcinogenicity endpoint was evaluated concerning hamster, mouse and rat species. The experimental results on the remaining species (cynomolgus, dog, rhesus) are insufficient and were discarded. When the same compound was tested in more than one specie, the experimental results for all species were stored in a single entry. Thus, for this study, the database was pre-processed in order to expand a single entry to produce an entry for each specie. This pre-processing resulted in a total of 2272 entries.

EPAFHM: The Fathead Minnow Acute Toxicity database was generated by the U.S. EPA Mid-Continental Ecology Division (MED) for the purpose of developing an expert system to predict acute toxicity from chemical structure based on mode of action considerations [14]. The database contains 614 organic chemicals, for the prediction of acute toxicity endpoint of environmental and industrial chemicals.

NCTRER: Researchers within FDA's National Center for Toxicological Research (NCTR) generated a database of experimental ER (estrogen receptor) binding results with the purpose of developing improved QSAR models to predict ER binding affinities. The NCTRER database provides activity classifications for a total of 224 chemical compounds, with a diverse set of natural, synthetic and environmental estrogens [17].

[1] http://www.epa.gov/ncct/dsstox/index.html, accessed Dec 2008.

DBPCAN: Water disinfection by-products database contains predicted estimates of carcinogenic potential for 178 chemicals.The goal is to provide informed estimates of carcinogenic potential to be used as one factor in ranking and prioritising future monitoring, testing, and research needs in the drinking water area [18].

The structures of chemicals in DSSTox are stored as SDF[2] files as well as SMILES[3] strings.

In addition to the original database entries, a number of 50 molecular descriptors was calculated with the GenerateMD software[4]. Molecular descriptors belong to a set of pre-defined categories [12,19]. In our data sets we used the constitutional-based descriptors, also known as *1D descriptors*, and the topological-based descriptors, also called *2D descriptors.*

Furthermore, we generated molecular *fingerprints* of type FP4[5] using Open Babel software [20]. These fingerprints were then converted into binary attributes meaning the presence or absence of a particular chemical substructure in the compound, such as, fused rings, alkene, lactone, enolether, in a total of 300 chemical substructures.

All the information was encoded in ARFF format which provides the necessary information to run the classification and regression algorithms. The classification data sets were obtained from the regression ones by establishing a threshold of 50% for the toxicity activity: below the 50 % the drug was not considered toxic and above that value it was considered toxic.

Table 1. Characterisation of the data sets. (a) Class distribution. (b) Number of Attributes.

data set	active	inactive
CPDB	1059	1213
EPAFHM	580	34
NCTRER	131	93
DBPCAN	80	98

(a)

Type of Features	Number of Attributes
1D	22
1D+FP	322
2D	564

(b)

Table 1(a) summarizes the four datasets used in this study. Except for the EPAFHM dataset all the datasets are balanced in terms of the number of active and inactive compounds. In all the experiments, we used the same number of attributes: 22 pure 1D descriptors, 300 molecular fingerprints, and 242 2D descriptors (Table 1(b)).

2.2 Machine Learning Methods

In this study we used classification and regression methods. The classification methods used cover most popular methods in Machine Learning and include

[2] Structure Data Format.

[3] Simplified Molecular Input Line Entry Specification.

[4] http://www.chemaxon.com, accessed Oct 2008.

[5] a set of SMARTS queries.

several variants of decision trees, Bayesian methods, instance base classification, and a variant of Support Vector Machines (SVMs). We experimented with different decision trees as they are widely considered one of the most interpretable methods, and interpretability may be of interest in this experiment. Regarding regression, we used approaches based in trees and based on SVMs.

All the machine learning methods used in our study are implemented in the Weka [21] software package. The classification methods used are enumerated in Table 2 and the list of regression methods used is enumerated in Table 3. The tree construction methods we have used are BF trees, CART, J48, ADTrees, and Functional Trees. We used Breiman's"random forests" as a popular methods that benefits from some of the understandability of trees and has been shown to achieve quite good accuracies. We also used two bayes methods: Naive Bayes is usually a good reference, and the "Bayes Network" method uses a greedy algorithm to construct the network. The K Star algorithm is a version of the famous $K-NN$ instance based classifier, and, last but not least, we used Weka's SMO SVM implementation.

We have used four regression methods: M5P and M5rules are based on model trees; in contrast, SMO is based on SVM technology, Linear Regression is a well known statistical method.

Table 2. Classification methods used in the present work

Method (Weka name)	Type
BF Tree	Decision Trees.
CART	Decision Trees
SVM SMO	Support Vector Machines
Naive Bayes	Bayesian classifier
Bayes Network	Bayesian classifier
K Star	Instance based classifier
AD Tree	Alternate Decision Trees
FT	Functional trees
J48	Decision tree
Random Forest (RF)	Ensemble method

Table 3. Regression methods used in the present work

Method (Weka name)	Type
M5P	Model Trees.
M5rules	Model Trees.
SVM SMO	Support vector machines
Linear Regression	statistical method

2.3 Experimental Design

The experiments were carried out on a cluster of 9 nodes where each node has two quad-core Xeon 2.4GHz and 32 GB of RAM and runs Linux Ubuntu 8.10. The Weka [21] version used was 3.6.0.

To estimate the predictive quality of the classification models we performed 10 fold cross-validation. The quality of the regression models were estimated measuring the Relative Absolute Errors (RAE):

$$RAE = \frac{\sum_{j=1}^{N} |P_{ij} - T_j|}{\sum_{j=1}^{N} |\overline{T_i} - T_j|}$$

in the formula, given the classifier i and the drug j, P_{ij} is the prediction made by i on j, $\overline{T_i}$ is the average value of actual activity over all drugs, and T_j is the actual activity value of drug j.

To handle missing values we considered three approaches: i) let the ML algorithm deal with the missing values; ii) use Weka pre-processing procedure for missing values and; iii) remove entries with missing values.

All ML algorithms were applied for each combination of missing values approach, and dataset (using 1D descriptors, 1D descriptors and Finger Prints (1D+FP) and all the information available).

3 Results and Discussion

3.1 Classification

Table 4 presents the average accuracy over 10 folds obtained by all classification methods considered, in all data sets and with different sets of features (1D, 1D+FP, and 1D+FP+2D)[6].

The last line in the table, for the ZeroR classifier, gives the baseline performance for the other learning methods. The ZeroR simply predicts the majority class in the training data. The values in bold are the best value for the column (for a data set and set of features). Overall, the classifiers are quite accurate on DBPCAN, where they achieve over 90% accuracy. All classifiers (except `bayesNet`) perform well in this dataset. EPAFHM is a very skewed dataset: predicting base class would give 95% accuracy. Even so, all classifiers exceed default accuracy, in fact the Functional Trees (FT) classifier achieves almost perfect accuracy. Performance is also quite good for NCTRER, between 77–86%, with best results for `ft` and `rf`. The hardest dataset was CPDBAS, where performance ranges from 55% for naive bayes (a little better than default class) up to 73% with random forests. Notice that in contrast to other datasets, nearest neighbor is quite effective in this dataset.

[6] The empty cells for NCTRER and BF tree and Cart are due to the incapacity of those two algorithms to handle missing values and all examples of the NCTRER data set have a missing value.

Table 4. Performance of classification algorithms. The values are obtained from the Weka package and are the average (over 10 folds) correctly classified instances obtained by different learning algorithms using 1D, 1D plus finger prints (1D+FP) plus 2D features (1D+FP+2D).

	CPDBAS			DBPCAN		
Algorithm	1D	1D+FP	1D+FP+2D	1D	1D+FP	1D+FP+2D
J48	66.4	64.0	68.7	89.9	93.3	91.6
BFTree	66.1	63.9	66.2	85.4	92.1	89.3
cart	65.6	66.7	69.1	85.9	91.0	90.6
ibk	62.1	59.6	62.4	91.0	92.7	91.0
SMO	58.4	60.9	65.9	92.8	92.7	92.1
Nbayes	57.8	58.0	54.4	87.1	89.3	83.7
bayesNet	57.7	56.9	58.4	81.5	88.2	84.3
kStar	70.8	68.9	72.0	88.2	92.7	89.3
ADTree	61.3	60.8	63.3	90.4	93.8	92.7
ft	66.9	70.3	70.2	92.7	94.4	95.5
rf	71.8	70.3	72.8	92.7	93.8	92.7
ZeroR	53.4	53.4	53.4	55.1	55.1	55.1
	EPAFHM			NCTRER		
Algorithm	1D	1D+FP	1D+FP+2D	1D	1D+FP	1D+FP+2D
J48	98.2	98.2	98.2	83.0	81.7	85.57
BFTree	97.9	97.9	97.5	-	-	-
cart	97.9	97.9	97.3	-	-	-
ibk	96.3	95.4	96.1	82.6	85.7	83.0
SMO	96.3	94.6	93.8	81.7	85.3	83.5
Nbayes	94.8	93.3	94.6	79.0	83.5	80.8
bayesNet	94.1	92.7	89.1	80.8	81.3	83.5
kStar	97.2	96.2	96.1	79.0	79.0	76.8
ADTree	98.5	98.2	98.2	83.9	81.7	84.4
ft	99.0	98.9	98.7	85.7	85.3	86.7
rf	98.7	97.6	97.2	83.9	85.3	87.1
ZeroR	94.5	94.5	94.5	58.5	58.5	58.5

Functional trees (FT) performed quite well on these datasets. FT tend to perform close to the best, and for the columns where FT performance is not the best, it is the second or third best. In all cases, the performance obtained by FT is better than the baseline.

Focusing on FT, the performance values for 1D descriptors in the four data sets clearly supports the hypothesis that 1D descriptors contain sufficient information for Machine Learning algorithms to construct accurate and simple predictive models to determine if a given molecule is toxic or not.

Extending the 1D set of descriptors to include fingerprints did not boost performance. However, extending the set of features to include the 2D set of descriptors improves the performance in most cases. The only exception is observed in the EPAFM data set. However, in this case, it was difficult to improve

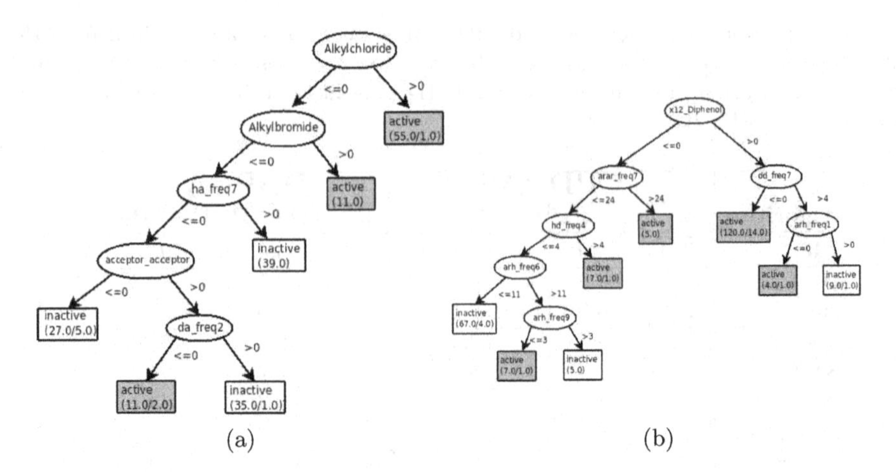

Fig. 1. Decision Tree to predict toxicity for the (a) DBPCAN and (b) NCTRER data sets

the performance obtained using 1D descriptors since it was already at 99%. Nevertheless, the results support the second hypothesis that states that extending the 1D set of descriptors of molecules with 2D descriptors improves the accuracy of the models constructed by Machine Learning algorithms to predict the degree of toxic activity of molecules.

An interesting question is whether the models discovered are supported or contribute to expertise in the area. Figure 1 shows two example trees obtained by the J48 algorithm. We chose J48 because it is particularly easy for an experiment to understand, and we chose two applications where J48 performs well.

The tree in Figure 1(a) presents a model for DBPCAN. The model is based on halogenated hydrocarbons (e.g., alkylchloride, alkylbromide), which are closely associated to toxicity. That such elements would be present in the model is unsurprising. First, it is well known that organisms have difficulty in metabolising halogens. Indeed the halogens such as fluorine, chlorine, bromine and iodine are elements avoided by medical chemists. Second, halogenated hydrocarbons (and aromatic rings such as the ones in diphenol) are hydrophobic. This property allows these molecules to cross biological membranes and deposit in the the fatty tissue, where they eventually can be involved in cancer process (a classical example is DDT).

More curious is that in this model the absence of halogenated hydrocarbons in "active" molecules seems to be compensated by having close-by hydrogen acceptors. (*acceptor_acceptor* > 0). Donor groups seem to relate with the functionality of drugs (and namely to specificity for certain targets). The existence of a large number of accepting groups with hydrogen bonds (more than 10), that seems to be contemplated in *acceptor_acceptor* > 10 may be associated with more promiscuous molecules, that is, with molecules connecting to a number of different targets with secondary effects.

Regarding the J48 model for NCTRER, the Diphenol group associated to molecules with a positive sharing coefficient between octanol and water seems to explain molecule activity/toxicity, very much in line with the previous discussion on hydrophobicity.

3.2 Regression

HR0: 1D descriptors contain sufficient information for Machine Learning algorithms to construct accurate and simple predictive models of the degree of toxic activity of molecules.

HR1: Extending the 1D set of descriptors of molecules with 2D descriptors improves the models constructed by Machine Learning algorithms to predict the degree of toxic activity of molecules.

The results of Table 5 show that the confirmation of HR0 depends on the data set. For EPAFHM the m5rules algorithm achieves a RAE of 26.6% that is quite good, whereas in CPDBAS data set the best score is 72%. A similar conclusion may be reached as HR1 is concerned. In CPDBAS data set we see a systematic reduction in RAE in all algorithms when the set of descriptors is enriched with 2D descriptors (with SVM algorithm it drops from 92.4% to 63.0%).

Table 5. Performance of the regression algorithms. Results obtained in the regression task by different learning algorithms using 1D, 1D plus finger prints (1D+FP) and 1D+FP plus 2D features (1D+FP+2D). Values represent RAE.

Algorithm	CPDBAS			DBPCAN			EPAFHM			NCTRER		
	1D	1D+FP	1D+FP+2D	1D	1D+FP	1D+FP+2D	1D	1D+FP	1D+FP+2D	1D	1D+FP	1D+FP+2D
m5p	73.6	70.8	67.5	41.5	42.4	45.6	34.6	33.9	36.9	50.2	**46.2**	56.2
m5rules	**72.4**	**68.8**	66.8	**39.3**	**41.6**	**44.4**	**26.6**	27.8	**24.3**	47.9	51.2	**46.1**
l.regression	95.0	90.9	82.9	52.5	53.4	99.2	110.1	122.1	147.5	61.5	63.1	100.0
svm smo	92.4	89.9	**63.0**	52.8	47.9	51.3	53.2	51.3	68.4	**43.1**	52.6	63.0

4 Related Work

In this study machine learning algorithms were applied to the task of predicting toxicity endpoints. Other solutions to the prediction toxicology problem have been published. In this section we report in related work. Notice that, to the best of our knowledge, we are the first the report extensive comparison results for these recent datasets.

In [22] decision tree models were constructed to select from a set of 20 attributes the ones whose values best discriminate a set of 904 rodent bioassay experiments. A classification system TIPT (Tree Induction for Predictive Toxicology) based on the tree was then applied and compared with neural networks models in terms of accuracy and understandability. The classification problem was also the subject of investigation in [23] where the Support vector machine (SVM) proved to be reliable in the classification of 190 narcotic pollutants (76 polar and 114 nonpolar). A selection algorithm was also used to identify three necessary attributes for the compounds discrimination and the leave-one-out cross-validation was the evaluated procedure. Again in [24] the problem is to predict potential toxicity of compounds depending on their physico-chemicals properties. It was used a wide variety of machine learning algorithms with Weka (machine learning software), including classical algorithms, such as k-nearest neighbours and decision trees, as well as support vector machines and artificial immune systems. SVMs proved to be the best algorithm followed by a neural network model. In [10] the ILP (Inductive logic programming) approach was used with support vector machines to extends the essentially qualitative ILP-based SAR to quantitative modelling. In this study a data set of 576 compounds with known fathead minnow fish toxicity was used and the results were compared with the commercial software TOPKAT. Furthermore, in [25] other machine learning approaches was analysed, such as Particle Swarm Optimisation (PSO) and Genetic Programming (GP), they are suitable for use with noisy, high dimensional data, as in commonly used in drug research studies. In [26] a literature review was done focus in predictive models such as partial-least square (PLS), support vector machines, neuronal nets, multiple linear regression and decision trees. A novel model to simulate complex chemical-toxicology data sets was reported in [27] and used to evaluate the performance of different machine learning approach, neuronal networks, k-nearest neighbours, linear discriminant analysis (LDA), naive Bayes, recursive partitioning and regression trees (RPART), and support vector machines.

5 Conclusions

The work reported in this paper addresses the problem of constructing predictive models of toxicity in a drug design setting. We have evaluated the usefulness of Machine Learning algorithms to construct accurate and simple models for toxicity. The study compared the usefulness of 1D and 2D molecular descriptors not only in the prediction of the degree of toxic activity but also the classification problem of predicting if a drug is toxic or not.

The results indicate that Machine Learning algorithms can effectively use 1D molecular descriptors to construct accurate and simple models to predict compound toxicity. The experiments also show that extending the set of 1D descriptors with 2D descriptors may improve the accuracy of the models, but that further work is required to take full advantage of these features.

Acknowledgements

This work has been partially supported by the project ILP-Web-Service (PTDC-/EIA/70841/2006) and by Fundação para a Ciência e Tecnologia. Nuno A. Fonseca is funded by FCT grant SFRH/BPD/26737/2006. Max Pereira is funded by FCT grant SFRH/BPD/37087/2007.

References

1. Plewczynski, D.: Tvscreen: Trend vector virtual screening of large commercial compounds collections. In: International Conference on Biocomputation, Bioinformatics, and Biomedical Technologies, BIOTECHNO 2008, pp. 59–63 (2008)
2. Graham, J., Page, C., Kamal, A.: Accelerating the drug design process through parallel inductive logic programming data mining. In: Computational Systems Bioinformatics Conference, p. 400. International IEEE Computer Society, Los Alamitos (2003)
3. Barrett, S.J., Langdon, W.B.: Advances in the Application of Machine Learning Techniques in Drug Discovery, Design and Development. In: Tiwari, A., Knowles, J., Avineri, E., Dahal, K., Roy, R. (eds.) Applications of Soft Computing: Recent Trends. Advances in Soft Computing, pp. 99–110. Springer, Heidelberg (2006)
4. Duch, W., Swaminathan, K., Meller, J.: Artificial intelligence approaches for rational drug design and discovery. Current Pharmaceutical Design 13, 1497–1508 (2007)
5. van de Waterbeemd, H., Gifford, E.: Admet in silico modelling: towards prediction paradise? Nat. Rev. Drug. Discov. 2(3), 192–204 (2003)
6. Neagu, D., Craciun, M., Stroia, S., Bumbaru, S.: Hybrid intelligent systems for predictive toxicology - a distributed approach. In: International Conference on Intelligent Systems Design and Applications, pp. 26–31 (2005)
7. Hansch, C., Maloney, P., Fujita, T., Muir, R.: Correlation of biological activity of phenoxyacetic acids with hammett substituent constants and partition coefficients. Nature 194, 178–180 (1962)
8. White, A., Mueller, R., Gallavan, R., Aaron, S., Wilson, A.: A multiple in silico program approach for the prediction of mutagenicity from chemical structure. Mutation Research/Genetic Toxicology and Environmental Mutagenesis 539, 77–89 (2003)
9. Richard, A.: Future of toxicology-predictive toxicology: An expanded view of "chemical toxicity". Chem. Res. Toxicol. 19(10), 1257–1262 (2006)
10. Amini, A., Muggleton, S., Lodhi, H., Sternberg, M.: A novel logic-based approach for quantitative toxicology prediction. J. Chem. Inf. Model. 47(3), 998–1006 (2007)
11. Dearden, J.: In silico prediction of drug toxicity. Journal of computer-aided molecular design 17(2-4), 119–127 (2003)
12. Ekins, S.: Computational Toxicology: Risk Assessment for Pharmaceutical and Environmental Chemicals. Wiley Series on Technologies for the Pharmaceutical Industry. Wiley-Interscience, Hoboken (2007)
13. Kazius, J., Mcguire, R., Bursi, R.: Derivation and validation of toxicophores for mutagenicity prediction. J. Med. Chem. 48(1), 312–320 (2005)
14. Russom, C., Bradbury, S., Broderius, S., Hammermeister, D., Drummond, R.: Predicting modes of toxic action from chemical structure: Acute toxicity in the fathead minnow (pimephales promelas). Environmental toxicology and chemistry 16(5), 948–967 (1997)
15. Richard, A., Williams, C.: Distributed structure-searchable toxicity (dsstox) public database network: a proposal. Mutation Research/Fundamental and Molecular Mechanisms of Mutagenesis 499, 27–52 (2002)
16. Gold, L., Manley, N., Slone, T., Ward, J.: Compendium of chemical carcinogens by target organ: Results of chronic bioassays in rats, mice, hamsters, dogs, and monkeys. Toxicologic Pathology 29(6), 639–652 (2001)

17. Fang, H., Tong, W., Shi, L., Blair, R., Perkins, R., Branham, W., Hass, B., Xie, Q., Dial, S., Moland, C., Sheehan, D.: Structure-activity relationships for a large diverse set of natural, synthetic, and environmental estrogens. Chem. Res. Toxicol. (14), 280–294 (2001)
18. Woo, Y., Lai, D., McLain, J., Manibusan, M., Dellarco, V.: Use of mechanism-based structure-activity relationships analysis in carcinogenic potential ranking for drinking water disinfection by-products. Environ. Health Perspect (110), 75–87 (2002)
19. Todeschini, R., Consonni, V., Mannhold, R., Kubinyi, H., Timmerman, H.: Handbook of Molecular Descriptors. Wiley-VCH, Chichester (2000)
20. Guha, R., Howard, M., Hutchison, G., Murray-Rust, P., Rzepa, H., Steinbeck, C., Wegner, J., Willighagen, E.: The blue obelisk – interoperability in chemical informatics. J. Chem. Inf. Model. 3(46), 991–998 (2006)
21. Witten, I.H., Frank, E.: Data Mining: Practical machine learning tools and techniques, 2nd edn. Morgan Kaufmann, San Francisco (2005)
22. Bahler, D., Stone, B., Wellington, C., Bristol, D.: Symbolic, neural, and bayesian machine learning models for predicting carcinogenicity of chemical compounds. J. Chemical Information and Computer Sciences 8, 906–914 (2000)
23. Ivanciuc, O.: Aquatic toxicity prediction for polar and nonpolar narcotic pollutants with support vector machines. Internet Electronic Journal of Molecular Design (2), 195–208 (2003)
24. Ivanciuc, O.: Weka machine learning for predicting the phospholipidosis inducing potential. Current Topics in Medicinal Chemistry (8) (2008)
25. Pugazhenthi, D., Rajagopalan, S.: Machine learning technique approaches in drug discovery, design and development. Information Technology Journal 5(6), 718–724 (2007)
26. Muster, W., Breidenbach, A., Fischer, H., Kirchner, S., Müller, L., Pähler, A.: Computational toxicology in drug development. Drug Discovery Today 8(7) (2008)
27. Judson, R., Elloumi, F., Setzer, R., Li, Z., Shah, I.: A comparison of machine learning algorithms for chemical toxicity classification using a simulated multi-scale data model. BMC Bioinformatics (2008)

Influence of Antigenic Mutations in Time Evolution of the Immune Memory – A Dynamic Modeling

Alexandre de Castro[1,2], Carlos Frederico Fronza[2], Poliana Fernanda Giachetto[1], and Domingos Alves[2,3]

[1] Embrapa Informática Agropecuária - EMBRAPA, CP 6041, Campinas, Brazil
acastro@cnptia.embrapa.br
[2] Dep. de Informática em Saúde, Universidade Federal de São Paulo – UNIFESP, CP 20266, São Paulo, Brazil
[3] Dep. de Medicina Social, Universidade de São Paulo – USP, CP 301, São Paulo, Brazil

Abstract. In this paper, we study the behavior of immune memory against antigenic mutation. Using a dynamic model proposed in previous studies, we have performed simulations of several inoculations, where in each virtual sample the viral population undergoes mutations. Our results suggest that the sustainability of the immunizations is dependent on viral variability and that the memory lifetimes are not random, what contradicts what was suggested by recent works. We show that what may cause an apparent random behavior of the immune memory is the antigenic variability.

Keywords: Dynamic model; Immune memory; Viral mutation; B-cell evolution; Antibodies.

1 Introduction

In recent decades, models have been widely used to describe biological systems, mainly to investigate global behaviors generated by cooperative and collective behavior of the components of these systems. More recently, several models were developed to study the dynamics of immune responses, with the purpose of comparing the results obtained by simulations, with the experimental results, so the connection between the existing theories on the functioning of the system and the available results can be adequately established [1,2,3,4,5].

From an immunological point of view, these models contribute to a better understanding of cooperative phenomena, as well as they lead to better understanding of the dynamics of systems out of equilibrium. Currently, the existing models to simulate the behavior of the immune system have been mainly based on differential equations, cellular automata and coupled maps. Processes and mechanisms of the natural immune system are being increasingly used for the

K.S. Guimarães, A. Panchenko, T.M. Przytycka (Eds.): BSB 2009, LNBI 5676, pp. 133–142, 2009.

development of new computational tools. However, the mathematical formalization of the functioning of the immune system is essential to reproduce, with computer systems, some of its key biological characteristics and skills, such as the ability of pattern recognition, information processing, adaptation, learning, memory, self-organization and cognition. Researchers, inspired by the intelligent techniques of recognition and elimination used by white blood cells, are planning a new generation of antivirus, with the purpose of searching in the biological systems the solutions to carry out strategic attacks, which may be the transforming elements of future technologies.

Within this scenario, in 2000, Lagreca et. al. [6] proposed a model that uses techniques of multi-spin coding and iterative solution of equations of evolution (coupled maps), allowing the global processing of a system of higher dimensions. The authors showed that the model is capable of storing the information of foreign antigens to which the immune system has been previously exposed. However, the results obtained by these authors for the temporal evolution of clones, include only the B cells, not taking into account the antibodies population soluble in the blood. In 2006, one of the present authors has proposed an extension of the Lagreca model, including the populations of antibodies. With this assumption, we considered not only the immunoglobulins attached to the surfaces of the B cells, but also the antibodies scattered in serum, that is, the temporal evolution of the populations of secreted antibodies is considered, to simulate the role in mediation of the global control of cell differentiation and of immunological memory [6,7,8,9]. In that work, our approach showed that the soluble antibodies alter the global properties of the network, diminishing the memory capacity of the system. According to our model, the absence or reduction of the production of antibodies favors the global maintenance of the immunizations. This result contrasts with the results obtained by Johansson and Lycke [10], who stated that the antibodies do not affect the maintenance of immunological memory.

Without considering terms of antigen mutation, this same extension [7,8,9] also led us to suggest a total randomicity for the memory lifetime, in order to maintain homeostasis of the system [11,12,13,14]. This random behavior was also recently proposed by Tarlinton et. al. [14]. In earlier work [7,8,9], the results indicated that, in order to keep the equilibrium of the immune system, some populations of memory B cells must vanish so that others are raised and this process seemed completely random. However, the results shown in this study suggest that the durability of immunological memory and the *raised-vanished* process is strongly dependent on the variability of viral populations. Thus, the lifetimes of immune memory populations are not random, only the antigenic variability from which they depend upon is a random feature, resulting in an apparent randomicity to the lifespan of B memory cells.

2 Extended Model and Methodology

In the model used here, the molecular receptors of B cells are represented by bitstrings with diversity of 2^B, where B is the number of bits in the string [15,16].

The individual components of the immune system represented in the model are the B cells, the antibodies and the antigens. The B cells (clones) are characterized by its surface receptor and modeled by a binary string. The epitopes – portions of an antigen that can be connected by the B cell receptor (BCR) – are also represented by bit-strings. The antibodies have receptors (paratopes) that are represented by the same bit-string that models the BCR of the B cell that produced them [16,17,18,19,20,21,22,23,24,25].

Each string shape is associated with an integer σ ($0 \leq \sigma \leq M = 2^B - 1$) that represents each of the clones, antigens or antibodies. The neighbors for a given σ are expressed by the Boolean function $\sigma_i = (2^i xor \sigma)$. The complementary form of σ is obtained by $\overline{\sigma} = M - \sigma$, and the temporal evolution of the concentrations of different populations of cells, antigens and antibodies is obtained as a function of the integer variables σ.

The equations that describe the behavior of clonal populations $y(\sigma, t)$ are calculated using an iterative process, for different initial parameters and conditions:

$$y(\sigma, t+1) = (1 - y(\sigma, t)) \times \left\{ m + (1-d)y(\sigma, t) + b\frac{y(\sigma, t)}{y_{tot}(t)}\zeta_{a_h}(\overline{\sigma}, t) \right\}$$

and all the complementary shapes included in the term $\zeta_{a_h}(\overline{\sigma}, t)$

$$\zeta_{a_h}(\overline{\sigma}, t) = (1 - a_h)[y(\overline{\sigma}, t) + y_F(\overline{\sigma}, t) + y_A(\overline{\sigma}, t)] + a_h \sum_{i=1}^{B} [y(\overline{\sigma}_i, t) + y_F(\overline{\sigma}_i, t) + y_A(\overline{\sigma}_i, t)], \quad (1)$$

In these equations, $y_A(\sigma, t)$ and $y_F(\sigma, t)$ are, respectively, the populations of antibodies and antigens, b is the rate of proliferation of B cells; $\overline{\sigma}$ and $\overline{\sigma}_i$ are the complementary forms of σ, and of the B nearest neighbors in the hypercube (with the ith bit inverted). The first term (m), within the brackets in equation (1) represents the production of cells by the bone marrow and is a stochastic variable. This term is small, but non zero. The second term in the bracket describes the populations that have survived to natural death (d), and the third term represents the clonal proliferation due to interaction with complementary forms (other clones, antigens or antibodies). The parameter a_h is the relative connectivity between a certain bit-string and the neighborhood of its image or a complementary form. When $a_h = 0.0$, only perfectly complementary forms are allowed. When $a_h = 0.5$, a string can equally recognize its image and its first neighbors. The factor $y_{tot}(t)$ is given by

$$y_{tot}(t) = \sum_{\sigma} [y(\sigma, t) + y_F(\sigma, t) + y_A(\sigma, t)]. \quad (2)$$

The temporal evolution of the antigens is determined by:

$$y_F(\sigma, t+1) = y_F(\sigma, t) - k\frac{y_F(\sigma, t)}{y_{tot}(t)} \times \{(1 - a_h)[y(\overline{\sigma}, t) + y_A(\overline{\sigma}, t)] + a_h \sum_{i=1}^{B} [y(\overline{\sigma}_i, t) + y_A(\overline{\sigma}_i, t)]\}, \quad (3)$$

where k is the rate with which populations of antigens or antibodies decay to zero. The population of antibodies is described by a group of 2^B variables, defined in a B-dimensional hypercube, interacting with the antigenic populations:

$$y_A(\sigma, t+1) = y_A(\sigma, t) + b_A \frac{y(\sigma, t)}{y_{tot}(t)} \times \left[(1 - a_h) y_F(\overline{\sigma}, t) + a_h \sum_{i=1}^{B} y_F(\overline{\sigma}_i, t) \right] - k \frac{y_A(\sigma, t)}{y_{tot}(t)} \zeta_{a_h}(\overline{\sigma}, t), \quad (4)$$

where the contribution of the complementary forms $\zeta_{a_h}(\overline{\sigma}, t)$ is again included in the final term, b_A is the rate of proliferation of antibodies, and k is the rate of removal of antibodies, which measures their interactions with other populations. The antibody populations $y_A(\sigma, t)$ (which represent the total number of antibodies) depend on the inoculated antigen dose. The factors $\frac{y_F(\sigma,t)}{y_{tot}(t)}$ and $\frac{y_A(\sigma,t)}{y_{tot}(t)}$ are responsible for the control and decay of the antigens and antibodies populations, while the factor $\frac{y(\sigma,t)}{y_{tot}(t)}$ is the corresponding accumulation factor for the populations of clones, in the formation of the immunological memory. The clonal population $y(\sigma, t)$ (normalized total number of clones) may vary from the value produced by bone marrow (m) to its maximum value (in our model, the unit), since the Verhulst factor limits its growth [26,27].

Eqs. (1)-(4) form a set of equations that describe the main interactions in the immune system between entities that interact through key-lock match, i.e., that specifically recognize each other. This set of equations was solved iteratively, considering the viral mutations as initial conditions.

Using virtual samples – representing hypothetically 10 individuals (mammals) with the same initial conditions – we inoculated *in silico* each one of the 10 samples with different viral populations (110, 250 and 350) with fixed concentration, where the virus strains occur at intervals of 1,000 time steps, that is, at each 1,000 time steps a new viral population, different from the preceding one, is injected in the sample. When a new antigen is introduced *in maquina*, its connections with all other entities of the system are obtained by a random number generator [7,8,9].

Changing the seed of the random number generator, the bits in the bit-strings are flipped and, taking into account that in our approach the bit-strings represent the antigenic variability, the bits changes represent, therefore, the corresponding viral mutations. Thus, in order to simulate the influence of viral mutation on the duration of immunological memory, the seed of the number generator is altered for each of the 10 samples. Figure 1(a) shows the design of the experiments. In this model, we use for the rate of apoptosis (d), or natural cell death, the value of 0.99, the proliferation rate of the clones is equal to 2.0, and of the antibodies is 100. The connectivity parameter a_h was considered equal to 0.01 and the term of bone marrow m was set to 10^{-7}. The value 0.1 was set to represent each virus strain (the antigen dose) and the length of the bit-string B was set to 12, corresponding to a potential repertoire of 4,096 distinct cells and

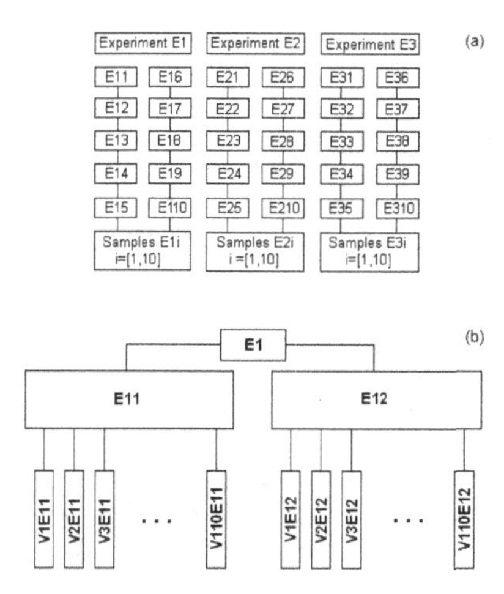

Fig. 1. Design of inoculations of viral populations in the samples for the experiments. In each experiment (a) were considered different lifetimes for the individuals. In (b), the design of inoculations of viral populations in the samples for the experiment E1 is shown.

receptors. Injections of different viral populations were administrated at time intervals corresponding to a period of the life of the individual.

3 Results

To investigate the relationship between the viral variability and the memory time of the population of lymphocytes that recognized a certain species of virus, three experiments *in silico* (E1, E2 and E3) were performed and organized as follows: in the first (E1), a lifetime equal to 110,000 was chosen for the individuals, in the second (E2), lifetime of 250,000 and in the third (E3), 350,000 time steps. In each of the experiments were used sets of 10 virtual samples (E1i, E2i and E3i), representing 10 identical individuals what, in our approach, corresponds to keep the amount of interaction of the coupled maps with the same initial conditions in all samples – 30 samples, taking into account the 3 experiments.

To simulate the viral strains, at each 1,000 time steps, a new dose of virus was administered. Therefore, in the first experiment were injected *in maquina* in the samples (individuals) 110, in the second, 250 and in the third 350 *distinct* viral populations and to represent the inoculation of *mutated* viral populations in each individual, the seed for the random number generator was changed for each one of the 30 samples. It is important to clarify that in our approach, *distinct* viral populations are populations of different species and *mutated* viral populations are genetic variations of the same population.

Figure 1(b) shows more clearly the entities used to simulate the behavior of memory against the antigenic variability. In the scheme, for example, the virus identified by V1E12 is a mutation of the virus V1E11 (both belong to the same original population, who suffered mutation) and the virus V2E11 is distinct of virus V1E11 (belonging to viral populations of different species). Figure 2 shows

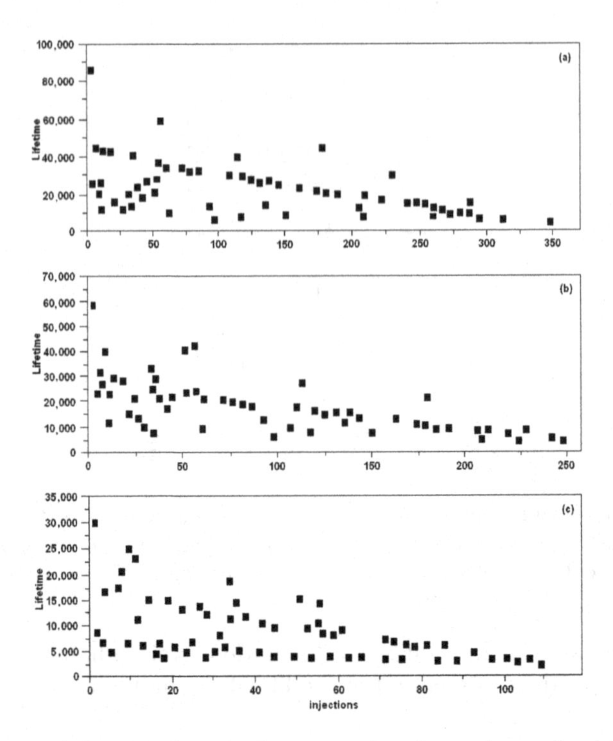

Fig. 2. Lifetime of the populations that recognize the antigens, for (a) 110, (b) 250 and (c) 350 inoculations

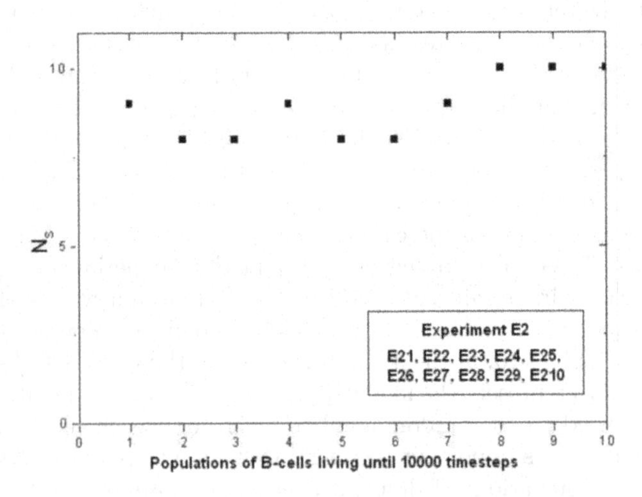

Fig. 3. Number of samples (N_S) of the experiment E2 with up to the tenth excited (live) clonal population, after 10,000 time steps have passed

the average lifetime of the populations of lymphocytes that specifically recognized the antigens in each of 3 experiments (E1, E2 and E3). In the experiments, the average lifetime of each population, excited by a kind of virus, is calculated over 10 samples (E1i, E2i and E3i). The difference in the average behavior of the memory in the three experiments (Figure 2(a), (b) and (c)) was expected, and is due to the fact that the samples were inoculated with virus of high genetic

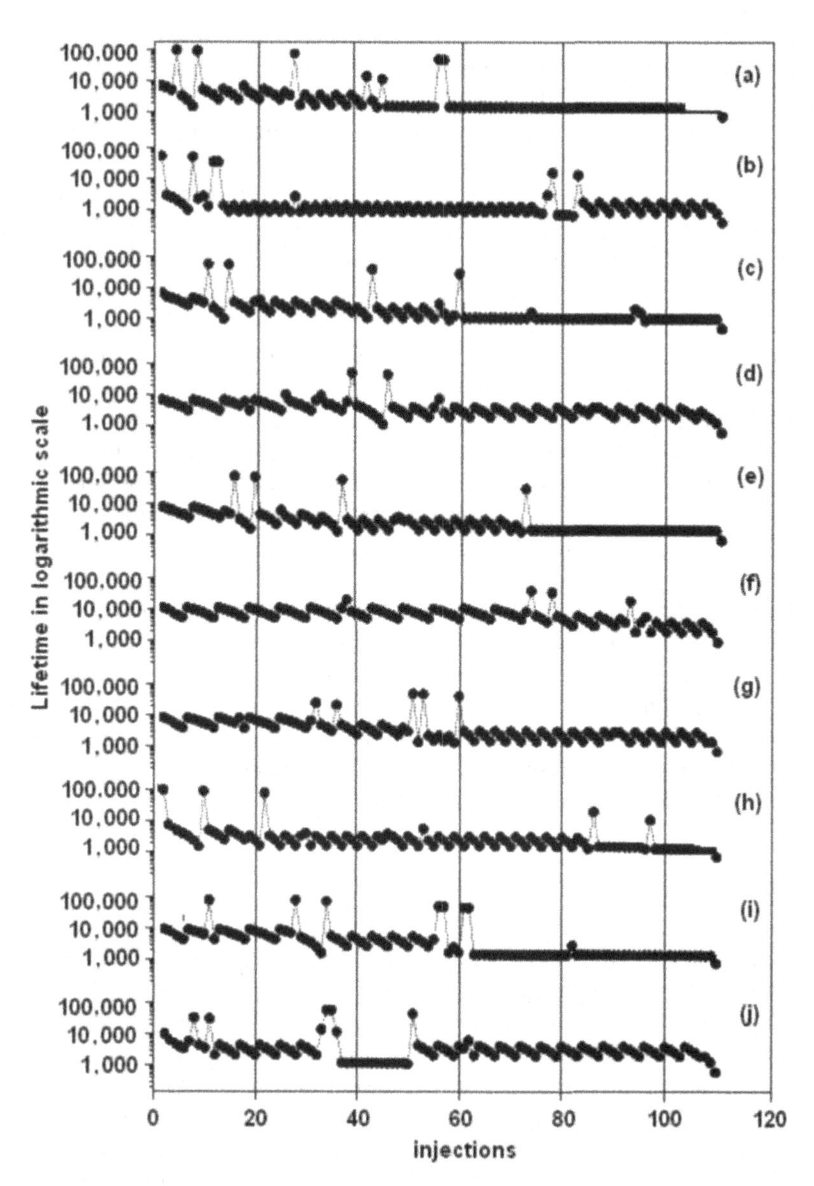

Fig. 4. Lifetime of the clonal populations in each sample (experiment E1). Similar behaviors were obtained for the experiments E2 and E3.

variability (in our approach, the random number generator seed was changed for each new sample to represent the genetic variability).

Even considering an uneven evolution for the memory in the three experiments, it is possible to see that there is a tendency of the first populations, on average, to survive longer than the subsequent. This result is consistent with the current hypothesis that, on average, the first vaccines administered in naïve individuals tend to have longer useful life than the vaccines administered in adulthood [28,29,30].

Figure 3 shows that for the experiment E2, the first clonal population (B cells population) remains alive in 9 samples, when the lifetime of individuals is 10,000. Similar situation was observed in experiment E3. However, in Figure 4(a) - (j) it is possible to visualize separately the behavior of each one of the 10 samples, when we administer, *in maquina*, 110 injections (experiment E1). From Figure 4, it is clearly noticed that we cannot say that the first clonal populations persist longer than the populations who later recognized other antigens, since the simulations indicate that only in two samples the first clonal population survived for a long period of time (Fig. 4(b) and (h)).

It is important to point out that this apparent discrepancy between the results of Figures 2 and 4 can be explained by the fact that in two samples the lifetime of the first excited clonal population was long, what determined that the simple arithmetic average was high – even if in the other samples the first clonal population has not survived for a long period of time. The latter result suggests that it is not safe to predict that the first excited populations tend to last longer – only on average these first immunizations tend to last longer than the others.

In specific situations, depending on the viral variability, our model shows that you cannot say that vaccines used in the first years of life of the individual provide with certainty immunization for long periods. Our results also suggest that the sustainability of immunization is dependent of the on viral variability and that the lifetimes of the memory populations are not completely random, but that the antigenic variability of which they depend on causes an apparent randomicity to the lifespan of memory lymphocytes.

4 Conclusion

We have performed simulations using a model of bit-strings that considers the temporal evolution of B cells, antigens and antibodies, and have studied the behavior of immune memory and which main factors are influencing the durability of vaccines or immunizations [7,8,9]. In previous studies, we have suggested that it is not possible to accurately determine the duration of memory [10], however, our results suggested that decreasing the production of antibodies, we can achieve greater durability for the immunizations [9]. Those simulations indicated that it is not possible to accurately determine the clonal populations that will survive for long periods, but those that survive, will have higher durability if there is a reduction in the production of soluble antibodies. This result may be biologically important, as it suggests a strategy to give more durability to the vaccines by inhibiting the production of antibodies harmful to the memory of the system.

In this article we present results that indicate that, besides the influence of populations of soluble antibodies [9], another factor that may be decisive for the durability of immunological memory is the antigenic mutation of the viral population, which brings on a reaction of the system. In this work we show that the lifetimes of the memory clones are not random, but the antigenic variability from which they depend originates an apparent randomicity to the lifespan of the B memory cells. As a consequence, our results indicate that the maintenance of the immune memory and its relation with the mutating antigens can mistakenly induce the *wrong deduction* of a stochasticity hypothesis for the sustainability of the immunizations, as proposed by Tarlinton et. al. [14]. Our results show that what presents a random aspect is the mutation of the viral species, resulting in an *apparent* unpredictable duration for the lifetime of the memory clones.

References

1. Ribeiro, L.C., Dickman, R., Bernardes, A.T.: 387, 6137 (2008)
2. Castiglione, F., Piccoli, B.: J. Theor. Biol. 247, 723 (2007)
3. Davies, M.N., Flower, D.R.: Drug Discov. Today 12, 389 (2007)
4. Sollner, J.: J. Mol. Recognit. 19, 209 (2006)
5. Murugan, N., Dai, Y.: Immunome Res. 1, 6 (2005)
6. Lagreca, M.C., de Almeida, R.M.C., dos Santos, R.M.Z.: Physica A 289, 42 (2000)
7. de Castro, A.: Simul. Mod. Pract. Theory 15, 831 (2007)
8. de Castro, A.: Eur. Phys. J. Appl. Phys. 33, 147 (2006)
9. de Castro, A.: Physica A 355, 408 (2005)
10. Johansson, M., Lycke, N.: Immunology 102, 199 (2001)
11. Roitt, I., Brostoff, J., Male, D.: Immunology, 5th edn. Mosby, New York (1998)
12. Abbas, A.K., Lichtman, A.H., Pober, J.S.: Cellular and Molecular Immunology, 2nd edn. W.B. Saunders Co., Philadelphia (2000)
13. Alberts, B., Bray, D., Lewis, J., Raff, M., Roberts, K., Watson, J.D.: Molecular Biology of the Cell, 4th edn. Garland Publishing (1997)
14. Tarlinton, D., Radbruch, A., Hiepe, F., Dorner, T.: Curr. Opin. Immunol. 20, 162 (2008)
15. Perelson, A.S., Weisbush, G.: Rev. Mod. Phys. 69, 1219 (1997)
16. Perelson, A.S., Hightower, R., Forrest, S.: Res. Immunol. 147, 202 (1996)
17. Playfair, J.H.L.: Immunology at a Glance, 6th edn. Blackwell Scientific Publications, Oxford (1996)
18. Levy, O.: Eur. J. Haematol. 56, 263 (1996)
19. Reth, M.: Immunol. Today 16, 310 (1995)
20. Moller, G.: Immunol. Rev. 153 (1996)
21. Jerne, N.K.: Clonal Selection in a Lymphocyte Network. In: Edelman, G.M. (ed.) Cellular Selection and Regulation in the Immune Response, pp. 39–48. Raven Press (1974)
22. Burnet, F.M.: The Clonal Selection Theory of Acquired Immunity. Vanderbuilt University, Nashville, TN (1959)

23. Jerne, N.K.: Ann. Immunol. 125, 373 (1974)
24. Celada, F., Seiden, P.E.: Immunol. Today 13, 53 (1992)
25. Celada, F., Seiden, P.E.: J. Theoret. Biol. 158, 329 (1992)
26. Verhulst, P.F.: Correspondance mathématique et physique 10, 113 (1838)
27. Verhulst, P.F.: Nouveaux Memoires de l'Academie Royale des Sciences et Belles-Lettres de Bruxelles 18, 1 (1845)
28. Gray, D., Kosko, M., Stockinger, B.: Int. Immunol. 3, 141 (1991)
29. Gray, D.: Ann. Rev. Immunol. 11, 49 (1993)
30. Mackay, C.R.: Adv. Immunol. 53, 217 (1993)

FReDD: Supporting Mining Strategies through a Flexible-Receptor Docking Database

Ana T. Winck, Karina S. Machado, Osmar Norberto-de-Souza, and Duncan D.D. Ruiz

PPGCC, Faculdade de Informática, PUCRS, Porto Alegre, RS, Brazil
{ana.winck,karina.machado,osmar.norberto,duncan.ruiz}@pucrs.br

Abstract. Among different alternatives to consider the receptor flexibility in molecular docking experiments we opt to execute a series of docking using receptor snapshots generated by molecular dynamics simulations. Our target is the InhA enzyme from *Mycobacterium tuberculosis* bound to NADH, TCL, PIF and ETH ligands. After testing some mining strategies on these data, we conclude that, to obtain better outcomes, the development of an organized repository is especially useful. Thus, we built a comprehensive and robust database called FReDD to store the InhA-ligand docking results. Using this database we concentrate efforts on data mining to explore the docking results in order to accelerate the identification of promising ligands against the InhA target.

Keywords: Molecular docking, database, data preparation, data mining.

1 Introduction

Advances in molecular biology and the intensive use of computer modeling and simulation tools have impacted the drug discovery process [1], turning viable the rational drug design (RDD) [2]. *In-silico* based RDD is a cycle based on the interaction between receptors and ligands obtained by molecular docking [3]. Most of the docking simulation software can deal with the ligand flexibility. Ligands are usually small molecules and their different conformations in the binding pocket are easily simulated [4]. However, limitations occur when it is necessary to consider the receptor flexibility, specially because of the large number of degrees of freedom accessible to a receptor macromolecule [4, 5].

The state of the art docking algorithms predict an incorrect binding pose for about 50-70% of all ligands when only a single fixed receptor conformation is considered [4]. Including the explicit flexibility of the receptor may help overcome this limitation [4-6]. In this work we tackle the receptor explicit flexibility by performing a series of molecular docking experiments considering, in each one of them, a different conformation (snapshot) of the receptor, generated by a molecular dynamics (MD) simulation [7]. As a result, a very large amount of data is generated, depending on the MD simulation time scale, which can increase exponentially. Our approach is to apply data mining strategies on these data, aiming to discover interesting patterns.

We consider as receptor the InhA (2-*trans*-enoyl ACP reductase) enzyme from *Mycobacterium tuberculosis* [8]. The explicit flexibility of the InhA enzyme was

K.S. Guimarães, A. Panchenko, T.M. Przytycka (Eds.): BSB 2009, LNBI 5676, pp. 143–146, 2009.

modeled based on a fully solvated MD simulation trajectory for 3,100 ps generated by the SANDER module of AMBER6.0 [9] as described in [10]. In this work we consider as ligands: nicotinamide adenine dinucleotide (NADH – 52 atoms) [8], triclosan (TCL – 18 atoms) [11], pentacyano(isoniazid) ferrate (PIF – 24 atoms) [12] and ethionamide (ETH – 13 atoms). The indicated number of atoms of each ligand refers to the total number of atoms after the ligand preparation. Docking experiments were performed by a proper workflow as described in [13]. For each ligand we submitted 3,100 docking experiments, one for each receptor snapshot from the MD trajectory. The workflow employs AutoDock3.05 [14] as docking software, using the simulated annealing protocol and 10 runs of execution (with rigid ligands).

We have already tested data mining techniques on results of molecular docking experiments [15]. However, the input files were produced with an exhaustive and manual effort. Having had this experience, we are convinced that to obtain more satisfactory outcomes, the development of an organized repository to store fully flexible-receptor docking results would be specially useful.

In this work we describe a comprehensive and robust database called FReDD (Flexible Receptor Docking Database) that is able to store the mentioned docking results in a detailed level. To the best of our knowledge there are neither published strategies involving the application of data mining algorithms nor facilities like FReDD to analyze docking results considering the explicit flexibility of receptors.

2 Developing FReDD Data Base: Towards Mining Experiments

FReDD database was developed aiming to store the complete docking results showing relations among its features. The data came from distinct sources such as each flexible-receptor molecular docking experiment, which involves characteristics of the atoms and residues of receptor and ligands. The final model of FReDD is composed of 16 tables shown in Figure 1. FReDD currently stores only data related to the InhA receptor and the four ligands. We have a total of 12,424,800 receptor atoms coordinates records. This number corresponds to 4,008 atoms of each snapshot multiplied by 3,100 snapshots. In addition, we get 3,248,330 records for the ligand atoms coordinates. Table 1 shows the current population of some FReDD tables.

FReDD's main benefit is to offer functionalities to quickly and efficiently provide inputs to be mined. Currently, we are applying mining techniques using WEKA [16] trying to figure out how to select snapshots having a given ligand. We decided to use as predictive attributes the minimum distances between the atoms in the ligand and the atoms in the receptor's residues, for each docking result, and the FEB value as a target attribute. These data is generated by a proper SQL statement handling 12,424,800 coordinate records of the receptor with 510,660 coordinate entries for the TCL ligand, for instance, producing more than 6 trillion of records for just one ligand. It is clear that this query is very difficult to be produced manually.

Having this query result, the next step is to assemble them in an input file with 271 attributes, in which the first two columns are the number of receptor snapshots and ligand conformations. The next 268 columns are the minimum distance found for each receptor residue, and the last column is the FEB value (the target attribute). For the TCL example, it was generated an input file with 28,370 records (which corresponds to the total number ligand conformations). The inputs provided have

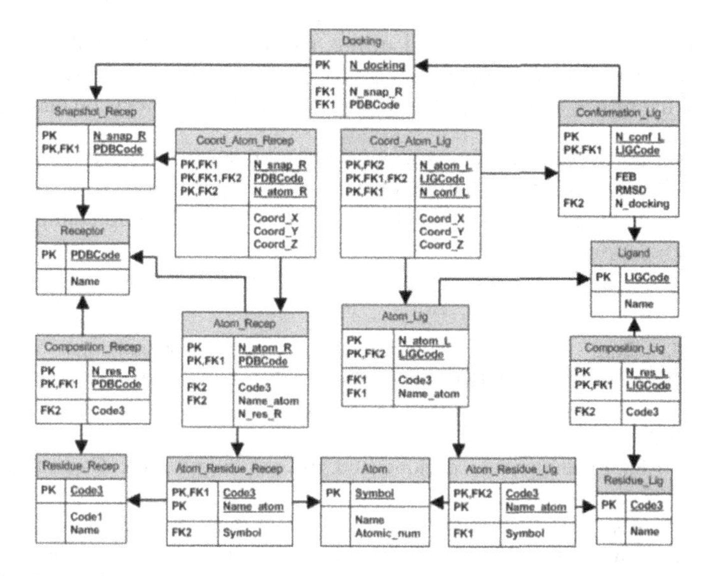

Fig. 1. The final model of FReDD, designed using the Microsoft Visio software

Table 1. MD snapshots and docking results stored in FreDD

Ligands	Number of atoms	Number of valid docking	Total of conformations	Total of coordinates
NADH	52	3,100	31,000	1,612,000
TCL	18	2,837	28,370	510,660
PIF	24	3,042	30,420	730,080
ETH	13	3,043	30,430	395,590
Total		12,022	120,220	3,248,330

been tested with classification and regression algorithms. Although we have performed several experiments which have produced promising results, they still have to be better analyzed. The report of this analysis is beyond the scope of this paper.

3 Final Considerations and Future Work

The InhA enzyme is an important target for the treatment of tuberculosis and is highly flexible [10]. Our docking experiments considered four distinct ligands: NADH, TCL, PIF and ETH. These experiments were performed taking into account a MD simulation time scale of 3,100 ps. Currently, we have a total of 12,022 valid docking results, 120,220 ligand conformations and 3,248,330 ligand coordinates. Our goal is to contribute to speed up the RDD process, developing ways to identify snapshots characteristics for their selection, as well as to indicate promising ligands to be further tested. Our approach is to evaluate these data through data mining techniques.

This paper main contribution is the development of the FReDD database. FReDD is able to store, index and retrieve docking results to easily produce mining inputs. Our tests so far have shown that this implementation contributed significantly to generate input files to be mined, as well as to ease data preparation. As future work, we intend to use the inputs generated and presented in this paper with different data mining techniques. Furthermore, based on the mined results, we intend to perform snapshots selection to check the results and analyze them in terms of accuracy, precision and quality.

Acknowledgements. This work was supported by grants from MCT/CNPq (14/2008 deliberative council). ATW is supported by CT-INFO/CNPq (17/2007 deliberative council) PhD scholarship. KSM is supported by a CAPES PhD scholarship.

References

1. Drews, J.: Drug discovery: A historical perspective computational methods for biomolecular docking. Current Opinion in Structural Biology 6, 402–406 (1996)
2. Kuntz, I.D.: Structure-based strategies for drug design and discovery. Science 257, 1078–1082 (1992)
3. Lybrand, T.P.: Ligand-protein docking and rational drug design. Current Opinion in Structural Biology 5, 224–228 (1995)
4. Totrov, M., Abagyan, R.: Flexible ligand docking to multiple receptor conformations: a pratical alternative. Current Opinion in Structural Biology 18, 178–184 (2008)
5. Cozzini, P., et al.: Target Flexibility: An Emerging Consideration in Drug Discovery and Design. Journal of Medicinal Chemistry 51, 6237–6255 (2008)
6. Alonso, H., et al.: Combining Docking and Molecular Dynamic Simulations in Drug Design. Med. Res. Rev. 26, 531–568 (2006)
7. van Gunsteren, W.F., Berendsen, H.J.C.: Computer Simulation of Molecular Dynamics Methodology, Applications and Perspectives in Chemistry. Angew. Chem. Int. Ed. Engl. 29, 992–1023 (1990)
8. Dessen, A., et al.: Crystal structure and function of the isoniazid target of Mycobacterium tuberculosis. Science 267, 1638–1641 (1995)
9. Pearlman, D.A., et al.: AMBER, a computer program for applying molecular mechanics, normal mode analysis, molecular dynamics and free energy calculations to elucidate the structures and energies of molecules. Comp. Phys. Commun. 91, 1–41 (1995)
10. Schroeder, E.K., et al.: Molecular Dynamics Simulation Studies of the Wild-Type, I21V, and I16T Mutants of Isoniazid-Resistant Mycobacterium tuberculosis Enoyl Reductase (InhA) in Complex with NADH: Toward the Understanding of NADH-InhA Different Affinities. Biophysical Journal 89, 876–884 (2005)
11. Kuo, M.R., et al.: Targeting tuberculosis and malaria through inhibition of enoyl reductase: compound activity and structural data. J. Biol. Chem. 278, 20851–20859 (2003)
12. Oliveira, J.S., et al.: An inorganic iron complex that inhibits wild-type and an isoniazid-resistant mutant 2-trans-enoyl-ACP (CoA) reductase from Mycobacterium tuberculosis. Chem. Comm. 3, 312–313 (2004)
13. Machado, K.S., et al.: Automating Molecular Docking with Explicit Receptor Flexibility Using Scientific Workflows. In: Sagot, M.-F., Walter, M.E.M.T. (eds.) BSB 2007. LNCS (LNBI), vol. 4643, pp. 1–11. Springer, Heidelberg (2007)
14. Morris, G.M., et al.: Automated Docking Using a Lamarckian Genetic Algorithm and Empirical Binding Free Energy Function. J. Comput. Chemistry 19, 1639–1662 (1998)
15. Machado, K.S., et al.: Extracting Information from Flexible Receptor-Flexible Ligand Docking Experiments. In: Bazzan, A.L.C., Craven, M., Martins, N.F. (eds.) BSB 2008. LNCS (LNBI), vol. 5167, pp. 104–114. Springer, Heidelberg (2008)
16. Waikato Environment for Knowledge, Analysis, `http://www.cs.waikato.ac.nz/ml/weka`

A Wide Antimicrobial Peptides Search Method Using Fuzzy Modeling

Fabiano C. Fernandes, William F. Porto, and Octavio L. Franco

Centro de Análises Proteômicas e Bioquímicas, Pós-Graduação em Ciências Genômicas e Biotecnologia, Universidade Católica de Brasília, Brasília - DF, Brazil, 70790-160
{fabianofernandesdf,williamfp7,ocfranco}@gmail.com

Abstract. The search for novel antimicrobial peptides in free databases is a key element to design new antibiotics. Their amino acid physicochemical features impact into the antimicrobial peptides activities. The relationship between the amino acid physicochemical properties and the antimicrobial target might have a fuzzy behavior. This study proposes a sequence similarity and physicochemical search method followed by a fuzzy inference system to find the most appropriated antimicrobial peptides for each domain. The proposed system was tested with NCBI's NR protein data file and the obtained peptide sub dataset will be tested *in vitro*.

Keywords: antimicrobial peptides, fuzzy modeling and drug design.

1 Introduction

Antimicrobial peptides (AMPs) are found in eukaryotes and they are used by the immune system to control bacterial infection [1]. A sequence similarity wide search for AMPs common patterns within protein sequences could be utilized to discover novel sequences that are useful for new drug design [2]. The physicochemical amino acids similarity may assist the search for common features in peptide sequence data. Moreover, investigation of common physicochemical rules in amino acid sequences with some similarities degree is important for elucidation of new sequences [2]. The antimicrobial activity can be affected by some intrinsic characteristics, such as peptide length, hidrophobicity and others [3]. In the present study, we developed a new method to search common patterns in protein sequences based on sequence similarity and amino acid physicochemical characteristics using fuzzy modeling. The Section 2 presents the proposed method. In Section 3 the proposed method is used to select the best putative sequences and Section 4 shows the conclusion.

2 Materials and Methods

The proposed method comprises the following steps: (i) select only the sequences with small size, between 6 and 59 amino acids residues from the NCBI's (National

K.S. Guimarães, A. Panchenko, T.M. Przytycka (Eds.): BSB 2009, LNBI 5676, pp. 147–150, 2009.

Center for Biotechnology Information) protein non redundant dataset (NR), (ii) based on a "seed" sequence and its suitable variation, select all matches within this subset (sequence similarity), (iii) calculate the physicochemical properties of all selected sequences, and (iv) using a Fuzzy Inference System, predict what are the most suitable primary structure. The search method will be complemented with *in vitro* testing to validate it. The Figure 1 shows the proposed method.

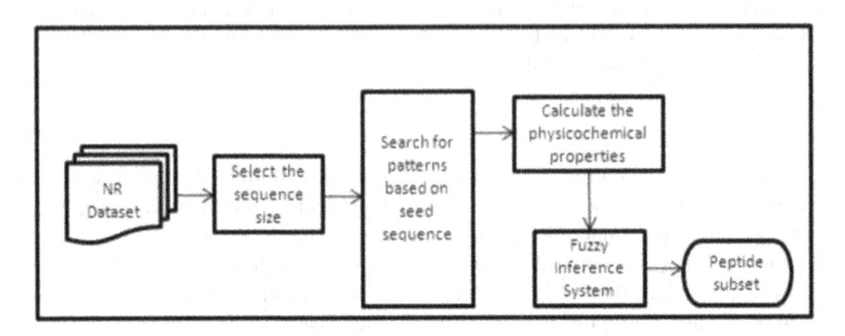

Fig. 1. Peptide search method using fuzzy modeling

Based on previous research, three novel antimicrobial peptides were isolated from green coconut water and used as a peptide "seed" and further used as a variation pattern filter [8]. After this filter, we searched for sequences with a length size earlier proposed obtaining a NR's file subset containing only the most suited AMPs candidates for this given "seed". After obtaining the NR's file subset with only the AMPs candidates, the average hydrophobicity of complete sequences were calculated in order to be used as a first physicochemical property. This property was chosen since hydrophobicity is an important attribute for antimicrobial peptide membrane interactions [5]. The second physicochemical property used was the hydrophobic to charged ratio residues sequences that can vary from 1:1 to 2:1 for adequate AMPs [3]. The Mathworks Matlab Fuzzy Inference System (FIS) [7] was used as a fuzzy modeling tool to find the most suited candidate for AMPs function. Firstly, we were not looking for the strongest candidates to AMPs, but prototypes that showed some weakness degree, leading some uniqueness sequences and structure. This action was carried out in order to filter obvious and conventional AMPs and reduce unspecific activities [5]. The FIS variables used

Table 1. Defined FIS rules

Rules
If Hydrophobicity is low *or* the Ratio is lower than 1 *Then* AMP is weak
If Hydrophobicity is medium *and* Ratio is adequate *Then* AMP is Specific
If Hydrophobicity is High *Then* Amp is strong

are Hydrophobicity, previously described Ratio and AMP property. The defined FIS rules are shown in Table 1.

The membership functions for hydrophobicity, ratio and AMPs are Gaussian MF, and Triangular MF. The surface plot for the FIS rules shows that the best residues sequences will have a degree of fuzziness between hydrophobicity and the ratio of hydrophobic to charged amino acids, as shown in Figure 2.

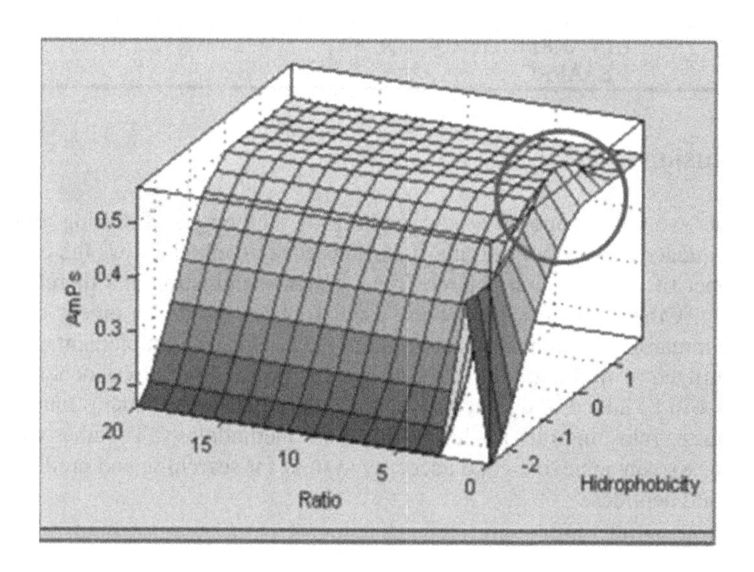

Fig. 2. The surface fuzziness between average hydrophobicity and the ratio of hydrophobic to charged amino acids. The circle shows the best peptides.

3 Results

In order to evaluate the proposed system accuracy, we used the "seed" SVAGRAQGM and the variations [A-Z][AVLIPMFWY][A-Z]G[KRH][A-Z][A-Z]G[A-Z] since it showed higher antibacterial activity when compared to TESYFVFSVGM and YC-SYTMEA and they are promising AMPs [8]. The seed variations were based on hydrophobicity common patterns and side chains [2]. The Table 2 shows a sample result of the first and second filtering scripts. From the original NR file with 7.153.872 amino acid residues sequences we obtained 1.510 amino acid residues sequences following the size between 6 to 59 and the "seed" variations. The calculations of hydrophobicity, the ratio of hydrophobic to charged amino acid residues sequences and the fuzzy output are also shown in Table 2. The FIS output variable was modeled to range from 0 indicating the putative non specific AMPs, to 0.6 indicating the probably most specific AMPs.

Table 2. A peptide sample of search for patterns and sequence size

gi	Sequence	Hydrophobicity	Ratio	Fuzzy
16304083	VDSVVGKEDGLGVENI HGSAAIASAYS	0.226	2.0	0.605
1432012	SIYLCSVDGRGTGELFF GEGSRLTVL	0.369	2.0	0.603
163733248	MLAVGRIGGNLNQIAQ WLNRAMLAGRTDLDA LTVARRMLTIERQLAQI VEAARRC	0.114	2.1	0.605

4 Conclusion

In this paper we have proposed a new method for searching and scoring amino acid residues sequences with AMPs similarities. This fuzzy method allows the choice of a small number of antimicrobial peptides for a better examination and therefore for *in vitro* testing. Other strategies have been used, such as the development of a novel peptide grammar to find unnatural active AMPs [1], which is complementary to strategy here utilized to find natural AMPs. In a near future, more physicochemical characteristics will be added in the fuzzy system, calibrate the membership functions and add more fuzzy rules, in order to clear validate our methodology. Together with the *in vitro* testing we can achieve a good accuracy system for searching and predicting new anti microbial peptides.

References

1. Loose, C., Jensen, K., Rigoutsos, I., Stephanopoulos, G.: A Linguistic Model for the Rational Design of Antimicrobial Peptides. Nature 443, 867–869 (2006)
2. Tomita, Y., Kato, R., Okochi, M., Honda, H.: A Motif Detection and Classification Method for Peptide Sequences Using Genetic Programming. J. Bios. Bioeng. 106, 154–161 (2008)
3. Brogden, A.K.: Antimicrobial Peptides: Pore Formers or Metabolic Inhibitors in Bacteria? Nature 3, 238–250 (2005)
4. National Center for Biotechnology Information, http://www.ncbi.nlm.nih.gov
5. Yeaman, M.R., Yount, N.Y.: Mechanisms of Antimicrobial Peptide Action and Resistance. Pharmacol Rev 55, 27–55 (2003)
6. Mount, D.W.: Bioinformatics: Sequence and Genome Analysis. Cold Spring Harbor Laboratory Press, New York (2000)
7. Mathworks. MatLab (2003), http://www.mathworks.com
8. Mandal, S.M., Dey, S., Mandal, M., Sarkar, S., Maria-Neto, S., Franco, O.L.: Identification and Structural Insights of Three Novel Antimicrobial Peptides Isolated from Green Coconut Water. Peptides 30, 633–637 (2008)

Identification of Proteins from Tuberculin Purified Protein Derivative (PPD) with Potential for TB Diagnosis Using Bioinformatics Analysis

Sibele Borsuk, Fabiana Kommling Seixas, Daniela Fernandes Ramos, Caroline Rizzi, and Odir Antonio Dellagostin*

Centro de Biotecnologia, Universidade Federal de Pelotas,
CP 354, 96010-900, Pelotas, RS, Brazil
Phone: +55 53 32757350
odirad@terra.com.br

Abstract. The PPD is widely used for diagnosis of bovine tuberculosis, however it lacks specificity and sensitivity, generally attributed the cross-reactions with antigens shared by other *Mycobacterium* species. These highlight the necessity of better diagnostic tools to detect tuberculosis in cattle. We have identified proteins present in PPD preparations (avium and bovine PPD) by LC-MS/MS (Liquid Chromatography/Mass Spectrometry/Mass Spectrometry). A total of 169 proteins were identified. From these, four PPD proteins identified in bovine PPD and absent in avium PPD (Mb1961c, Mb2898, Mb2900 and Mb2965c) were select for bioinformatics analysis to identify an antigen with potential for TB diagnosis. We identified Mb1961c, a secreted protein, which has low identity proteins found in environmental mycobacteria. This protein could be used for the development of a diagnostic test or a synthetic PPD for bovine tuberculosis.

Keywords: bioinformatics analysis, LC-MS/MS, bovine tuberculosis, diagnostic.

1 Introdution

Tuberculosis (TB) continues to be a worldwide problem for both humans and animals [1]. The fact that current tuberculosis control in cattle is entirely based on the tuberculin skin test to identify infected animals, and the subsequent slaughter of such tuberculin-positive animals, limits the use of BCG vaccination, an important measure that could aid in the control bovine tuberculosis [2]. However, the tuberculin skin test cannot distinguish between a *M. tuberculosis/bovis* infection and exposure to environmental mycobacteria. These cross-reactions are generally attributed to the presence in PPD of antigens shared by other *Mycobacterium* species [3][4]. Thus, the identification of immunogenic proteins in

* Corresponding author.

K.S. Guimarães, A. Panchenko, T.M. Przytycka (Eds.): BSB 2009, LNBI 5676, pp. 151–155, 2009.

PPD that are not present in environmental mycobacteria or with low homology with *M. bovis* proteins could be useful for the development of specific diagnostic tests [5]. The active components of PPD had not been determined yet [6][7]. In order to identify the proteins that are present in bovine and avium PPD preparations we used LC-MS/MS (Liquid Chromatography/Mass Spectrometry/Mass Spectrometry). Identification of proteins from PPD preparations and *in silico* analysis were used to identify potential antigens that could allow the development of more specific and sensitive tests.

2 Material and Methods

2.1 PPD Sample

Bovine and avian PPD samples were obtained from the Instituto Biolgico, So Paulo, Brazil and from the Veterinary Laboratories Agency, Weybridge, Surrey, UK.

2.2 1-D Gel Electrophoresis, Sample Digestion and LC-ESI-MS/MS

Fifty microlitres of PPD preparation (avium or bovine) were mixed with 25μL of SDS loading buffer and boiled for 5 min prior to separation on a 10 cm long, 1 mm thick 18% SDS-polyacrylamide gel. Sliced gel bands (1 mm thick) were washed twice with 50% acetonitrile (ACN) in 25 mM ammonium bicarbonate (NH_4HCO_3). The gel pieces were dehydrated by incubating them with 50 mL 100% ACN. Proteins were reduced using 10 mM DTT and alkylated with 55 mM iodoacetamide (IAA); both in 25 mM NH_4HCO_3. The gel pieces were dehydrated with ACN as described above, and rehydrated in 25 mM NH_4HCO_3 containing 0.01 mg/mL modified trypsin (Promega). Proteins were digested by trypsin for 16-20h at 37 °C. Then, the tryptic peptides were eluted by incubating the gel pieces with 50μL 1% trifluoroacetic acid (TFA). The tryptic peptides were and concentrated to 10μL using Speed-Vac®(Eppendorf). The samples were analysed by nano-LC-ESI-MS/MS using an Ultimate 3000 nano-HPLC (Dionex, Amsterdam) coupled to a QSTARTM XL (Applied Biosystems, Foster city, USA), a hybrid quadropole time of flight instrument. The peptides were separated on a 75μm, 15 cm, C18 pepmap column (Dionex-LC packing).

2.3 *In Silico* Analysis

The peak lists of MS/MS spectra were submitted to MASCOT (v2.2.04, Matrix Science, London, UK) and analysed using the MS/MS ion search program. Protein identifications were made by searching against the National Center for Biotechnology Information non-redundant (NCBInr) data base with the taxonomy set for the *M. tuberculosis complex*. An automatic decoy database search was performed for each sample. http://genolist.pasteur.fr/BoviList/index.html was used to convert the NCBI nr accession numbers to Mb numbers. The tools available at http://www.cbs.dtu.dk and http://tbsp.phri.org websites were used to determinate the protein class. The alignment of selected proteins with other mycobacteria was made using the protein sequence in FASTA format. This sequence was obtained directly from the National Center for Biotechnology Information (NCBI) using Protein BLAST search. Settings were *Mycobacterium* Organism and PSI-BLAST Algorithm. The ClustalW2 (http://www.ebi.ac.uk/Tools/clustalw2/index.html) was used to produce sequence alignments.

3 Results and Discussion

Four different PPD preparations, two avium and two bovine produced in Brazil and in UK were submitted to LC/MS/MS. A total of 169 proteins among the four PPD samples were identified by LC-MS/MS analysis by our group (unpublished data). From these we selected four potential proteins (Mb1961c, Mb2898, Mb2900 and Mb2965c), which were identified as present only in bovine PPD preparations and were secreted proteins. All the proteins are secreted with Sec signal peptide and are involved in cell wall and cell processes. We also chose these antigens because the *M. bovis* secreted antigens are more frequently recognised by immune system than other proteins [8]. Homology among the Mb1961c, Mb2898, Mb2900 and Mb2965c and genes found in other mycobacterial species, especially environmental mycobacteria, were analysed using the tools in the ClustalW2 website. Identity lower than 50% was not considered for analysis. The table 1 summarize the results from the analysis.

Table 1. Results of alignment of the Mb1961c, Mb2898, Mb2900 and Mb2965c protein sequences using Protein BLAST

Proteins	kDa	Description	Access Number	Species	Identity (%)
Mb1961c	16504	immunogenic protein MPT63	gi:27065400	*M. tuberculosis H37Rv (NP_217391.1)*	98
				M. marinum (YP 0018511341)	85
				M. smegmatis (YP_889653.1)	54
Mb2898	24412	lipoprotein MPT83	gi: 166786205	*M. tuberculosis H37Rv (NP_217391.1)*	100
				M. canetti (ABV487571)	99
				M. kansasii (ABV487491)	76
Mb2900	8674	immunogenic protein MPT70	gi:6469704	*M. tuberculosis H37Rv (NP_217391.1)*	67
				M. kansasii (ABV487511)	76
				M. marinum (YP 0018511381)	67
				M. smegmatis (YP_884484.1)	71
Mb2965c	24296	mycocerosic acid synthase membrane-associated MAS protein	gi:15610077	*M. tuberculosis H37Rv (NP_217391.1)*	99
				M. bovis BCG Pasteur (YP 9790461)	99
				M. bovis BCG Tokyo(YP 0026460003)	99
				M. leprae (NP 301233)	87
				M. ulcerans (YP 905920)	78
				M. marinum(YP 001850071)	78
				M. avium (YP 880564)	61

Although we have selected the Mb2965c as one of the potential antigen for diagnosis, the *in silico* analysis shows that it matched with the vaccinal strain and with several environmental mycobacteria associated to bovine, which could result in antigenic cross-reactivity. Synthetic peptides derived from Mb2898 (MPT83) have been tested alone or in combination with other antigens, with limited success [9][10]. We identified in PPD samples a potential secreted protein, Mb1961c, which has orthologous sequence only in *M. tuberculosis* H37Rv, *M.smegmatis* and *M. marinum*. This is important for the development of a test able to detect animals with bovine TB. Homology with a *M. marinum* and *M. smegmatis* protein is not a concern, as the first on is a free-living environmental mycobacteria, which has not been implicated with bovine infection [11]. MPT63 was identified as a potent antigenic target for the humoral immune response in experimental bovine TB [12][13]. Therefore, the antigen could be used in a diagnostic test alone as a synthetic PPD or in combination with other antigens previously tested [9][10][14] increasing the specificity of TB diagnostic tests.

In conclusion, proteomics and *in silico* analysis was successfully used to identify potential antigens present in PPD samples. It allowed the identification of antigens that could possibly be used in TB specific diagnostic tests.

References

1. WHO: Global Tuberculosis Control. WHO report. WHO/CDS/TB/2002 (2002)
2. Hewinson, R.G., Vordermeier, H.M., Buddle, B.M.: Use of the bovine model of tuberculosis for the development of improved vaccines and diagnostics. Tuberculosis (Edinb) 83(1-3), 119–130 (2003)
3. Daniel, T.M., Janicki, B.W.: Mycobacterial antigens: a review of their isolation, chemistry, and immunological properties. Microbiol. Rev. 42(1), 84–113 (1978)
4. Young, D.B.: Heat-shock proteins: immunity and autoimmunity. Curr. Opin. Immunol. 4(4), 396–400 (1992)
5. Vordermeier, H.M., Chambers, M.A., Buddle, B.M., Pollock, J.M., Hewinson, R.G.: Progress in the development of vaccines and diagnostic reagents to control tuberculosis in cattle. Vet. J. 171(2), 229–244 (2006)
6. Klausen, J., Magnusson, M., Andersen, A.B., Koch, C.: Characterization of purified protein derivative of tuberculin by use of monoclonal antibodies: isolation of a delayed-type hypersensitivity reactive component from M. tuberculosis culture filtrate. Scand. J. Immunol. 40(3), 345–349 (1994)
7. Rowland, S.S., Ruckert, J.L., Cummings, P.J.: Low molecular mass protein patterns in mycobacterial culture filtrates and purified protein derivatives. FEMS Immunol. Med. Microbiol. 23(1), 21–25 (1999)
8. ANON.: The development of improved tests for the diagnosis of Mycobacterium bovis infection in cattle. Report of the Independent Scientific Group Vaccine Scoping Sub-Committee. Defra Publications, London (2006)
9. Vordermeier, H.M., Cockle, P.C., Whelan, A., et al.: Development of diagnostic reagents to differentiate between Mycobacterium bovis BCG vaccination and M. bovis infection in cattle. Clin. Diagn. Lab Immunol. 6(5), 675–682 (1999)

10. Vordermeier, H.M., Whelan, A., Cockle, P.J., Farrant, L., Palmer, N., Hewinson, R.G.: Use of synthetic peptides derived from the antigens ESAT-6 and CFP-10 for differential diagnosis of bovine tuberculosis in cattle. Clin. Diagn. Lab Immunol. 8(3), 571–578 (2001)
11. Gluckman, S.J.: Mycobacterium marinum. Clin. Dermatol. 13(3), 273–276 (1995)
12. Lyashchenko, K.P., Pollock, J.M., Colangeli, R., Gennaro, M.L.: Diversity of antigen recognition by serum antibodies in experimental bovine tuberculosis. Infect. Immun. 66(11), 5344–5349 (1998)
13. Manca, C., Lyashchenko, K., Wiker, H.G., Usai, D., Colangeli, R., Gennaro, M.L.: Molecular cloning, purification, and serological characterization of MPT63, a novel antigen secreted by Mycobacterium tuberculosis. Infect. Immun. 65(1), 16–23 (1997)
14. Buddle, B.M., Parlane, N.A., Keen, D.L., et al.: Differentiation between Mycobacterium bovis BCG-vaccinated and M. bovis-infected cattle by using recombinant mycobacterial antigens. Clin. Diagn. Lab Immunol. 6(1), 1–5 (1999)

Mapping HIV-1 Subtype C gp120 Epitopes Using a Bioinformatic Approach

Dennis Maletich Junqueira[1], Rúbia Marília de Medeiros[1], Sabrina Esteves de Matos Almeida[1], Vanessa Rodrigues Paixão-Cortez[2], Paulo Michel Roehe[3], and Fernando Rosado Spilki[4]

[1] Centro de Desenvolvimento Científico e Tecnológico (CDCT), Fundação Estadual de Produção e Pesquisa em Saúde (FEPPS), Porto Alegre/RS, Brazil
maletich@ig.com.br
[2] Departamento de Genética, Instituto de Biociências, Universidade Federal do Rio Grande do Sul, Porto Alegre/RS, Brazil
[3] Virology Laboratory, Instituto de Pesquisas Veterinárias Desidério Finamor (CPVDF) – FEPAGRO Animal Health
[4] Instituto de Ciências da Saúde, Centro Universitário FEEVALE, Novo Hamburgo/RS , Brazil

Abstract. Human Immunodeficiency Type-1 subtype C (HIV-1C) is rapidly diverging among populations causing more than 48% of infections worldwide. HIV-1C gp120's 128 sequences available at Genbank were aligned and submitted to phylogenetic analysis. Three major clusters were identified: 72 sequences aligned with a Brazilian 0072eference sequence; 44 sequences aligned with an Ethiopian sequence and 12 could be group along with Indian isolates. A search was made for conserved HIV-1C cytotoxic T lymphocyte (CTL) epitopes to A*0201, A*0301, A*1101 e B*07 human leukocyte antigen (HLA) alleles (using Epijen software). Five most conserved epitopes were recognized: QMHEDIISL, CTHGIKPVV, NLTNNVKTI, AITQACPKV, CTRPNNNTR. Our results showed a recognized evolutionary force of HIV-1 to escape from CTL responses mutating sites that can be negatively select by host's immune system. The present study brings up a new approach to *in silico* epitope analysis taking into account geographical informations on virus diversity and host genetic background.

Keywords: HIV-1, subtype C, bioinformatics, epitope, gp120, HLA.

1 Introduction

Human immunodeficiency virus type 1 (HIV-1) is the etiological agent of Acquired Immunodeficiency Syndrome (AIDS) [1]. Subtype C is the most prevalent HIV-1 (HIV-1C) form and it is responsible for more than 48% of infections worldwide [2]. It has also been reported as the main cause of infections in India and in several countries in Africa [2]. Despite the high prevalence of subtype B in South America [3] in Southern Brazil, subtype C also circulates as a predominant form in infections due to recent introduction of this viral genotype in the region [4]. Recent data points to regional clusters within subtypes, showing significant differences in subtype C among

K.S. Guimarães, A. Panchenko, T.M. Przytycka (Eds.): BSB 2009, LNBI 5676, pp. 156–159, 2009.

isolates from different parts of the world [5]. Such data is important for research on HIV vaccines, since the heterogeneity in viral populations is a major drawback for the development of an effective vaccine [6].

A putative CTL epitope screening on HIV proteins using bioinformatics tools might examine whether diverse HLA alleles in a given population would be more likely to generate an immune response to any particular HIV subtype. The present study aims to predict putative CTL epitopes on gp120 of HIV-1C on the four HLA alleles most prevalent in the world's population.

2 Materials and Methods

2.1 Sequence Selection

All sequences analyzed in this study were selected from the GenBank database. Three hundred and twenty *gp120* sequences of HIV-1C were selected and examined. Sequences smaller than 1300 base pairs (bp) in length were excluded from this study.

Selected sequences were analyzed with Rega HIV Subtyping Tool version 2.0 [7] for confirmation of subtypes and with Stanford database [8] to check the presence of premature stop codons. Sequences not subtyped as C and/or presenting premature stop codons were discarded.

Sequences were aligned with HIV-1C *gp120* reference sequences from different parts of the world (collected from Los Alamos Data Base) on the ClustalX 2.0 software [9]. All sequences were edited and verified manually using Bioedit [10].

2.2 Construction of Phylogenetic Trees

Neighbor-joining phylogenetic trees were constructed under HKY distance matrix implemented in PAUP 4.0 software [11] to analyze the relationship among the sequences that best satisfied the selected criteria. The phylogenetic trees were used to group isolates from different parts of the world with similar genetic characteristics into clades.

2.3 T Cell Epitope Prediction

T cell epitope mapping was performed on selected sequences using Epijen online software [12]. Four HLA alleles were selected in Epijen to generate putative epitopes: HLA A*0201, HLA A*0301, HLA A*1101 and HLA B*07. Peptides recognized by these four alleles give rise to a putative epitope recognition in more than 90% of the global population, regardless of ethnicity [13]. The NetChop online software [14] was used to reiterate proteasome cleavage sites and to select epitopes correctly processed.

3 Results

Following the established criteria for peptide screening, 128 different sequences representing HIV-1C gp120 were selected for further analysis to predict T cell epitopes. Phylogenetic analysis grouped the examined sequences into three major clades:

i) *Brazilian*, comprising 72 sequences that aligned with sequence U52953 (reference strain of Brazilian origin); ii) *Ethiopian*, comprising 44 sequences that aligned with U46016 (reference obtained from Ethiopian subjects); and iii) *Indian*, which included 12 isolates which aligned with Indian reference strain AF067155.

The Epijen software detected 140 different gp120 epitopes recognizable by HLA A*0201, HLA A*0301, HLA A*1101 and HLA B*07 alleles. All epitopes comprised nine amino acids; and at least one epitope was identified in each of the 128 gp120 sequences by at least one of the four HLA alleles. HLA A*0201 was the only allele to recognize epitopes in all 128 *gp120* sequences.

Five peptides (QMHEDIISL, CTHGIKPVV, NLTNNVKTI, AITQACPKV, CTRPNNNTR) were the most conserved epitopes within all 128 HIV-1C gp120 sequences examined, being recognized by three HLA alleles: HLA A*0201, HLA A*0301 and HLA A*1101. Regarding HLA B*07 recognized a high number of non conserved epitopes (93%). In addition, no conserved epitope among Brazilian, Ethiopian and Indian clades could be detected.

4 Discussion

The multitude of variants generated by the interplay between the immune response to HIV-1C gp120 antigens and different host genotypes was glanced in this study. The predictions on HLA A*0201 show that this allele is capable of recognizing a diversity of epitopes on HIV-1C gp120, particularly those on Ethiopian HIV-1 isolates. On the other hand, HIV-1C gp120 sequences from Indian cluster displayed a few number of epitopes potentially recognizeable by the four HLA alleles examined in this study. Such limited reactivity might be a bias generated by the scarce number of sequences of such region available in GenBank, versus the limited number of HLA alleles examined here.

Previous reports pointed out a few HLA A2 epitopes to HIV-1 and suggested that the virus evolved in such a way as to escape immune recognition by the most common HLA alleles found in humans [6], [15]. The results shown here revealed that the highest number of identical epitopes were recognized by HLA A*0201 allele out of the four alleles examined (HLA A*0201, HLA A* 0301, HLA A*1101, HLA B*07). Despite the predicted recognition of identical epitopes in a high number of sequences to HLA A*0201, it presented 70% of non-conserved epitopes, reflecting a high propensity of HIV-1 to mutate and escape CTL responses [6].

The conserved QMHEDIISL and its variants (diverging from it in one or two amino acids) were recognized in 98% of sequences by HLA A*0201 allele. This high level of conservation may be attributable to the structural importance of this site to the protein function.

The HLA B*07 supertype has been shown to be associated with high viral loads and rapid progression to disease. A recent study found that HLA B7-restricted epitopes were conserved in as many as 93% of the analyzed sequences analyzing several subtypes and viral genes [6]. The results obtained here shown a few number of conserved epitopes to this supertype due to the specific predictions to HIV-1C *gp120*.

The present study gives a new aproach to *in silico* epitope analysis. It is expected that the identification of epitopes as carried out here may provide some help towards the development of an effective HIV vaccine.

References

1. McMichael, A.J., Rowland-Jones, S.L.: Cellular Immune Responses to HIV. Nature 410, 980–987 (2001)
2. Novitsky, V.A., Montano, M.A., McLane, M.F., et al.: Molecular cloning and phylogenetic analysis of human immunodeficiency virus type I subtype C: a set of 23 full-length clones from Botswana. J. Virol. 73, 4427–4432 (1999)
3. Russell, K.L., Carcamo, C., Watts, D.M., et al.: Emerging genetic diversity of HIV-1 in South America. AIDS 14, 1785–1791 (2000)
4. Bello, G., Passaes, C.P., Guimaraes, M.L., Lorete, R.S., et al.: Origin and evolutionary history of HIV-1 subtype C in Brazil. AIDS 22, 1993–2000 (2008)
5. WHO-UNAIDS Report from a meeting Vaccine Advisory Comitee Geneva. Approaches to the development of broadly protective HIV vaccines: challenges posed by the genetic, biological and antigenic variability of HIV-1. AIDS 15, W1–W25 (2001)
6. De Groot, A.S., Jesdale, B., Martin, W., et al.: Mapping cross-clade HIV-1 vaccine epitopes using a bioinformatics approach. Vaccine 21, 4486–4504 (2003)
7. Oliveira, T., Deforche, K., Cassol, S., et al.: An Automated Genotyping System for Analysis of HIV-1 and other Microbial Sequences. Bioinfomatics 21(19), 3797–3800 (2005)
8. Rhee, S.Y., Gonzales, M.J., Kantor, R., et al.: Human immunodeficiency virus reverse transcriptase and protease sequence database. Nucleic Acids Res. 31, 298–303 (2003)
9. Thompson, J.D., Gibson, T.J., Plewniak, F., et al.: The ClustalX windows interface: flexible strategies for multiple sequence alignment aided by quality analysis tools. Nucleic Acids Res. 24, 4876–4882 (1997)
10. Hall, T.A.: BioEdit: a user-friendly biological sequence alignment editor and analysis program for Windows 95/98/NT. In: Nucl. Acids Symp. Ser., vol. 41, pp. 95–98 (1999)
11. Swofford, D.L.: PAUP*: phylogenetic analysis using parsimony (*and other methods) Sinauer Associates, Sunderland, Mass. (2001)
12. Doytchinova, I.A., Guan, P., Flower, D.R.: EpiJen: a server for multi-step T cell epitope prediction. BMC Bioinformatics 7, 131 (2006)
13. Sette, A., Sidney, J.: Nine major HLA class I supertypes account for the vast preponderance of HLA-A and -B polymorphism. Immunogenetics 50, 201–212 (1999)
14. Kesmir, C., Nussbaum, A., Schild, H., Detours, V., Brunak, S.: Prediction of proteasome cleavage motifs by neural networks. Prot. Eng. 15(4), 287–296 (2002)
15. Shankar, P., Fabry, J.A., Fong, D.M., et al.: Three regions of HIV-1 gp160 contain clusters of immunodominant CTL epitopes. Immunol. Lett. 52, 23–30 (1996)

MHC: Peptide Analysis: Implications on the Immunogenicity of Hantaviruses' N protein

Maurício Menegatti Rigo, Dinler Amaral Antunes, Gustavo Fioravanti Vieira, and José Artur Bogo Chies

Núcleo de Bioinformática, Department of Genetics, UFRGS, Porto Alegre, Brazil
{mauriciomr1985,dinler}@gmail.com, gusfioravanti@yahoo.com.br, jabchies@terra.com.br

Abstract. Hantaviruses, members of the *Bunyaviridae* family, are enveloped negative-stranded RNA viruses with tripartite genomes – S, M and L. The S genome codes for a nucleocapsid (N) protein, which is quite conserved among different species of the hantavirus genus and possess a recognized immunogenicity. In this work we analyzed the sequence of two regions in this protein ($N_{94\text{-}101}$ and $N_{180\text{-}188}$), which presents T cell epitopes for two species of hantaviruses – Sin Nombre and Puumala. Interestingly, the same region has no described epitopes for Hantaan virus, despite its similarity. A study using a bioinformatic approach for the construction of MHC:peptide complexes was performed to detect any variation on the cleft region that could explain such degrees of immunogenicity. Our results shown topological and charges differences among the constructed complexes.

Keywords: MHC-I, epitopes, Hantavirus, molecular docking.

1 Introduction

As members of the *Bunyaviridae* family, hantaviruses have a tripartite ssRNA(–) genome coding for a RNA-dependent RNA polymerase, two glycoproteins which are inserted into the viral envelope membrane, and the N protein associated with the viral genome [1]. The hantavirus nucleocapsid (N) protein fulfills several key roles in virus replication and assembly [1]. Also, it presents cross-reactivity among different members of the hantavirus genus [2]. Considering the recognized immunogenicity of this protein and that the final objective of our work is the development of a vaccine, two aspects should be here considered: the analysis of antigenic processing pathway and the presentation of epitopes to T cell on the MHC class I context. In a previous work (personal communication) we observed a concordance between conserved regions in the N protein and epitopes of hantaviruses described in literature [2]. Maeda *et al.* described epitopes for Puumala virus (PUU) and Sin Nombre virus (SNV) at region $N_{94\text{-}101}$ and $N_{180\text{-}188}$, however epitopes for Hantaan virus (HTN) in the same region were not observed, although the high similarity between these sequences (75% and 67%, respectively). Additionally, *in silico* simulations showed that epitopes of HTN

K.S. Guimarães, A. Panchenko, T.M. Przytycka (Eds.): BSB 2009, LNBI 5676, pp. 160–163, 2009.

in these regions can be generated by the antigen processing pathway in the same way as PUU and SNV. This data suggests a strong influence from some amino acids of the epitopes on the TCR recognition and on immune response induction. Therefore, the construction of a MHC:peptide complex for topological and charge analysis of these regions is quite important.

In the present work we analyzed two specific regions of the N protein (94-101/180-188) from three different species of hantaviruses – PUU, SNV and HTN. This analysis was made on modeled MHC:peptide complexes, mainly on the TCR surface contact, searching for charges and topological differences that could explain the different levels of immunogenicity observed among the epitopes from different hantaviruses.

2 Material and Methods

The epitope peptide sequences related to N_{94-101} and $N_{180-188}$ region from PUU, SNV and HTN were obtained from literature and written in the FASTA format. The crystal structure of murine MHC alleles, H2-K^b (1G7P) and H2-D^b (1WBY), were used to modeling the complexes of interest. Each peptide sequence was fitted on the parameters of the specific MHC allele epitope pattern using the SPDBV program [3]. Since we have the amino acids sequences within the parameters of three-dimensional shape, energy minimization was performed with the GROMACS 3.3.3 program [4] to stabilize the molecule, simulating an aqueous solution. The next step was a construction of the MHC:peptide complex using AutoDock 4.0 program [5]. After the construction of the first complex, a second energy minimization was performed aiming to a better interaction of the side chains of the MHC with the epitope. The MHC was separated from its epitope and a second docking was carried out for the construction of the final MHC:peptide complex. The surface of the resulting complex was visualized with the GRASP2 program [6] (Figure 1), where the electrostatic charge distribution and the shape were analyzed.

3 Results

All FASTA sequences were perfect fitted with PDB allele-specific peptides. Information about the positions of the anchor residues for H2-K^b and H2-D^b was obtained from SYFPEITHI [7], a MHC-ligand databank. Epitopes described in literature for SNV and PUU at the studied regions have the same sequence, therefore they were analyzed as a unique epitope. After two rounds of molecular docking and energy minimization we found good values of binding energy (BE) for each region (Table 1). At the topological and charge levels, categorical differences were found in MHC:peptide complex for N_{94-101} region, on the discordant residues. These differences could be important in TCR recognition. The $N_{180-188}$ region showed almost none differences at charge distribution, but a topological discrepancy was verified.

Table 1. Best binding energy values for each region of Hantavirus species

Hantavirus specie	Protein/Region	Peptide Sequence*	MHC allele	Best BE-1st docking (Kcal/mol)	Best BE - 2nd docking (Kcal/mol)
Hantaan	N_{94-101}	SMLSYGNV	H2-K^b	-6,11	-6,48
Sin Nombre & Puumala	N_{94-101}	SSLRYGNV	H2-K^b	-4,98	-7,49
Hantaan	$N_{180-188}$	SLPNAQSSM	H2-D^b	-11,07	-14,73
Sin Nombre & Puumala	$N_{180-188}$	SMPTAQSTM	H2-D^b	-12,01	-14,55

*Anchor amino acids are underlined.

4 Discussion

The described epitopes for N_{94-101} region are presented by H2-K^b allele, while $N_{180-188}$ are presented by H2-D^b allele [2]. There are a difference of only two amino acids at N_{94-101} region between SNV/PUU and HTN; at $N_{180-188}$ region, the difference is 3 amino acids. These substitutions seems to affect the recognition by the immune system, since there are no epitopes described for hantavirus in that region [2].

A first approach analyzed the binding energy value of molecular docking with peptides from these two regions of SNV/PUU and HTN. The computational program used was AutoDock. Values for binding energy were all negative. There are no data in literature about the best value for binding energy of peptide and MHC. However, this value is directly proportional to the entropy of the system. Thus, we admitted that

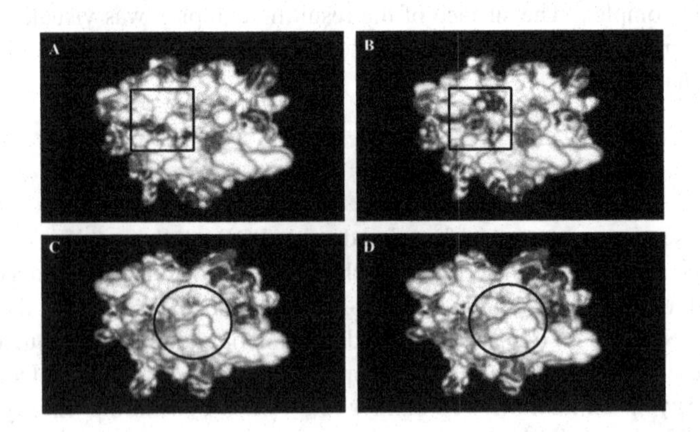

Fig. 1. Charge distribution of MHC:peptide complexes (Top-view) visualized at GRASP program. Peptide sequences from HTN (A) and SNV/PUU (B) on H2-Kb allele in N94-101 region; and from HTN (C) and SNV/PUU (D) on H2-Db allele in N180-188 region. Black squares show charge and topological differences between the first two complexes. Black circles show only topological difference on the remaining complexes.

the lowest value should be the best value. Accordingly to Binding energy data, all peptides have a good potential for attachment to the MHC. The conservation of anchor amino acids among the sequences for each MHC restriction could explain these good binding values.

It is known that changes in the distribution of charges interfere with the TCR recognition [8]. The best MHC:peptide models were visualized and analyzed at GRASP program, which provides a molecular surface with charge distribution, through the adjust of electric potential (Figure 1). The dark regions represent charged zones (both negative or positive charges). We could observe a topological and, especially, a charge difference between the MHC:peptide complexes of $N_{94\text{-}101}$ region, mainly on the fourth (Ser/Arg) residue. The $N_{180\text{-}188}$ region showed only topological changes, and the charges seems to remain equally distributed in both complexes. Our results showed that more than just charge differences, topological differences could explain the abrogation of an immune response.

References

1. Tischler, N.D., Rosemblatt, M., Valenzuela, P.D.T.: Characterization of Cross-reactive and Serotype-specific Epitopes on the Nucleocapsid Proteins of Hantaviruses. Virus Res. 135, 1–9 (2008)
2. Maeda, K., West, K., Toyosaki-Maeda, T., Rothman, A.L.: Identification and Analysis for Cross-reactivity Among Hantaviruses of H-2b-restricted Cytotoxic T-lymphocyte Epitopes in Sin Nombre Virus Nucleocapsid Protein. J. Gen. Virol. 85, 1909–1919 (2004)
3. Guex, N., Peitsch, M.C.: SWISS-MODEL and the Swiss-PdbViewer: an Environment for Ccomparative Protein Modeling. Electrophoresis 18, 2714–2723 (1997)
4. Van Der Spoel, D., Lindahl, E., Hess, B., Groenhof, G., Mark, A.E., Berendsen, H.J.: GROMACS: Fast, Flexible, and Free. J. Comput. Chem. 26, 1701–1718 (2005)
5. Morris, G.M., Goodsell, D.S., Halliday, R.S., Huey, R., Hart, W.E., Belew, R.K., Olson, A.J.: Automated Docking Using a Lamarckian Genetic Algorithm and an Empirical Binding Free Energy Function. J. Comput. Chem. 19, 1639–1662 (1998)
6. Petrey, D., Honig, B.: GRASP2: Visualization, Surface Properties, and Electrosttics of Macromolecular Structures and Sequences. Meth. Enzymol. 374, 492–509 (2003)
7. Rammensee, H., Bachmann, J., Emmerich, N.N., Bachor, O.A., Stevanovic, S.: SYFPEITHI: Database for MHC Ligands and Peptide motifs. Immunogenetics 50, 213–219 (1999)
8. Kessels, H.W., de Visser, K.E., Tirion, F.H., Coccoris, M., Kruisbeek, A.M., Schumacher, T.N.: The Impact of Self-tolerance on the Polyclonal CD8+ T-cell Repertoire. J. Immunol. 172, 2324–2331 (2004)

An Ontology to Integrate Transcriptomics and Interatomics Data Involved in Gene Pathways of Genome Stability

Giovani Rubert Librelotto[1], José Carlos Mombach[1], Marialva Sinigaglia[2], Éder Simão[1], Heleno Borges Cabral[3], and Mauro A.A. Castro[2]

[1] UFSM – Universidade Federal de Santa Maria.
Av. Roraima, 1000, Santa Maria - RS, 97105-900, Brasil
librelotto@inf.ufsm.br, jcmombach@smail.ufsm.br,
eder.simao@terra.com.br

[2] UFRGS – Universidade Federal do Rio Grande do Sul
Porto Alegre, Brasil
megsinigaglia@yahoo.com.br, mauro@ufrgs.br

[3] UNIFRA, Centro Universitário Franciscano
Santa Maria - RS, 97010-032, Brasil
hc@eiconet.com.br

Abstract. Disruption of the mechanisms that regulate cell-cycle checkpoints, DNA repair, and apoptosis results in genomic instability and the development of cancer. The description of the complex network of these pathways requires news tools to integrate large quantities of experimental data in the design of biological information systems. In this context we present Ontocancro an extensible ontology and an architecture designed to facilitate the integration of data originating from different public databases in a single- and well-documented relational database. Ontocancro is an Ontology stored in a knowledge database designed to be a source of information to integrate transcriptomics and interatomics data involved in gene pathways of Genome Maintenance Mechanisms (GMM).

1 Introduction

Genome maintenance mechanisms are shown to be critical for cell homeostasis since their malfunctioning can predispose to cancer. Repair, apoptosis and chromosome stability pathways comprise the cornerstone of GMM [CMdAM07]. The information about these pathways are disseminated in various databases as NCI-Nature, BioCarta, KEGG, Reactome, Prosite, GO and others. Ontocancro was created with the intention of integrating the information of genes involved in GMM from several curated databases. This data integration is difficult for biological data lack a unified vocabulary and need constant update what is provided by Ontocancro. Additionally, it allows the integration of transcriptome data provided by some Affymetrix microarrays platforms with interactome data from the STRING database, which has information about protein interactions.

This paper describes the integration of data from biological information system using the ontology paradigm, in order to integrate transcriptomics and interatomics data involved in gene pathways of genome stability and to generate an homogeneous view

K.S. Guimarães, A. Panchenko, T.M. Przytycka (Eds.): BSB 2009, LNBI 5676, pp. 164–167, 2009.

of those resources. It is organized as follows. Ontologies and Topic Maps are presented in section 2. The proposed ontology is described section 3. section 4 presents the concluding remarks.

2 Ontologies and Topic Maps

An ontology is a way of describing a shared common understanding, about the kind of objects and relationships which are being talked about, so that communication can happen between people and application systems [Gua98]. In other words, it is the terminology of a domain (it defines the universe of discourse).

An ontology consists of a set of axioms which place constraints on sets of individuals ("classes") and the types of relationships permitted among them. These axioms provide semantics by allowing systems to infer additional information based on the data explicitly provided. The data described by an ontology is interpreted as a set of "individuals" and a set of "property assertions" which relate these individuals to each other.

Topic Maps are abstract structures that can encode knowledge and connect this encoded knowledge to relevant information resources [PH03]. Topic Maps allow a domain knowledge representation in semantic networks, composed of topics and associations. In fact topic maps would indeed gain effectiveness and interoperability either through explicit formalization of ontologies specifically built and dedicated for topic map control, or through declaration of commitment to pre-defined ontologies, not specifically designed for that use.

Topic Maps take the main concepts described in the information resources and relates them together independently of what is said about them in the information being indexed. This means managing the meaning of the information, rather than just the information because topic maps take a step back from the details and focusing on the forest rather than the trees.

3 Ontocancro

Ontocancro is an ontology stored in a knowledge database designed to be a source of information to integrate transcriptomics and interatomics data involved in gene pathways of genome stability.

In pratice this data integration is difficult for biological data are disseminated in various databases, lack a unified vocabulary, and need constant update. Ontocancro aims to facilitate this task through integration of information on genome stability pathways from curated databases as NCI-Nature, BioCarta, KEGG, Reactome, Prosite, GO, and also from databases on physical interactions among proteins as STRING.

Due to a lack of consensus in the gene set definition of each different pathway in the databases and in the scientific literature, we developed the Ontocancro pathways that include all documented genes (from databases and literature) listed in these pathways [CDM+08]. Ontocancro pathways are meant to be a comprehensive gene lists for each pathway. The expanded Ontocancro pathways include, additionally, the genes that interact physically with these genes (interactions determined from the STRING database at 0.9 confidence level).

3.1 Ontocancro's Architecture

Daily, a lot of data is stored into biological databases and its organization requires an integrated view of their heterogeneous information systems. In this situation, there is a need for an approach that extracts the information from their data sources and fuses it in a semantic network. Usually, this is achieved either by extracting data and loading it into a central repository that does the integration before analysis, or by merging the information extracted separately from each resource into a central knowledge base. This is the Ontocancro's goal.

To achieve this aim, we propose the architecture described in figure 1. We developed a parser to extract the data from biological databases and store them into several XML files [Bra02]. These generated files are stored into a XML database called eXist.

Using this database, Metamorphosis [LRH$^+$08] links the information resources to a single ontology according to the Topic Maps format. We make Topic Maps a key component of the new generation of Web-aware knowledge management solutions. In this way, Metamorphosis let us achieve the semantic interoperability among heterogeneous information systems because the relevant data, according to the desired information specified through an ontology, is extracted and stored in a topic map.

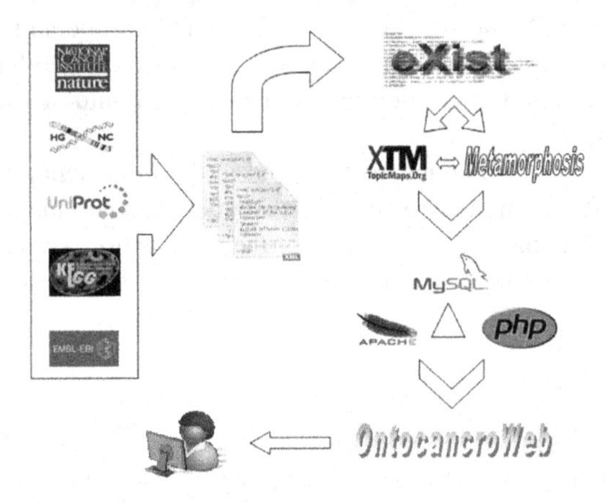

Fig. 1. Ontocancro' Architecture

A MySQL database contains the generated Topic Map and allows its processing by PHP scripts running over an Apache web server. This web server publishes the Ontocancro web page. Thus the navigation over the topic map is led by a semantic network and provides an homogeneous view over the resources. The Ontocancro pathways can be browsed in *http://www.ontocancro.org/*.

4 Conclusion

This paper described the integration of data from biological information systems using the ontology paradigm, in order to integrate transcriptomics and interatomics data involved in gene pathways of genome stability.

Due to a lack of consensus in the gene set definition of each different pathway in the databases and in the scientific literature, we developed the Ontocancro pathways that are meant to be comprehensive gene lists for each pathway above integrating a vast number of biological databases. The gene lists include sets of identifiers that allow unique identification of each gene in Ontocancro in the different databases. Some identifiers include EntrezGene ID, Ensembl Gene ID; Approved Symbol and Approved Name provided by HGNC; Unigene; and microarray identifiers from Affymetrix.

This proposal uses Metamorphosis for the automatic construction of Topic Maps with data extracted from the various biological data sources, and a semantic browser to navigate among the information resources. Topic Maps are a good solution to organize concepts, and the relationships between those concepts, because they follow a standard notation – ISO/IEC 13250 – for interchangeable knowledge representation. With this methodology the original information resources are kept unchanged and we can have as many different interfaces to access it as we want.

The ontology is being expanded to include other genetic pathways whose involvement in GMM are documented in the literature and databases. It will also provide interoperability with ViaComplex, a free software developed to visualize landscape maps of gene expression networks. ViaComplex can be found in:
http://lief.if.ufrgs.br/pub/biosoftwares/viacomplex/

Acknowledgements. This work has been partially supported by Brazilian Agency CNPq (grant 478432/2008-9). We acknowledge NCI-Nature, BioCarta, KEGG, Reactome, Prosite, GO, and STRING databases for providing public access to their data.

References

[Bra02] Bradley, N.: The XML Companion, 3rd edn. Addison-Wesley, Reading (2002)

[CDM+08] Castro, M.A.A., Dalmolin, R.L.S., Moreira, J.C.F., Mombach, J.C.M., de Almeida, R.M.C.: Evolutionary origins of human apoptosis and genome-stability gene networks. Nucleic Acids Res. 36(19), 6269–6283 (2008)

[CMdAM07] Castro, M.A.A., Mombach, J.C.M., de Almeida, R.M.C., Moreira, J.C.F.: Impaired expression of ner gene network in sporadic solid tumors. Nucleic Acids Res. 35(6), 1859–1867 (2007)

[Gua98] Guarino, N.: Formal Ontology and Information Systems. In: Conference on Formal Ontology (FOIS 1998) (1998)

[LRH+08] Librelotto, G.R., Ramalho, J.C., Henriques, P.R., Gassen, J.B., Turchetti, R.C.: A framework to specify, extract and manage Topic Maps driven by ontology. In: SIGDOC 2008. ACM Press, New York (2008)

[PH03] Park, J., Hunting, S.: XML Topic Maps: Creating and Using Topic Maps for the Web. Addison-Wesley, Reading (2003)

Author Index

Almeida, Sabrina Esteves de Matos 156
Alves, Domingos 133
Antunes, Dinler Amaral 160
Arrial, Roberto T. 73
Astrovskaya, Irina 1

Bazzan, Ana L.C. 86
Berger, Pedro A. 73
Borsuk, Sibele 151
Brigido, Marcelo M. 73
Brito, Rui M.M. 121

Cabral, Heleno Borges 164
Camacho, Rui 121
Carballido, Jessica Andrea 36
Carvalho, André C.P.L.F. de 109
Castro, Alexandre de 133
Castro, Mauro A.A. 164
Cerri, Ricardo 109
Chies, José Artur Bogo 160
Cliquet, Freddy 24
Costa, Ivan G. 48
Costa, Vítor Santos 121

Delbem, Alexandre C.B. 97
Dellagostin, Odir Antonio 151
Dias, Ulisses 13
Dias, Zanoni 13

Fernandes, Fabiano C. 147
Fertin, Guillaume 24
Fonseca, Nuno A. 121
Franco, Octavio L. 147
Fronza, Carlos Frederico 133

Gabriel, Paulo H.R. 97
Gallo, Cristian Andrés 36
Giachetto, Poliana Fernanda 133
Gómez, Juan Carlos 60

Junqueira, Dennis Maletich 156

Larese, Mónica G. 60
Lemke, Ney 86
Librelotto, Giovani Rubert 164
Lorena, Ana C. 48

Machado, Karina S. 143
Medeiros, Rúbia Marília de 156
Mombach, José Carlos 164

Norberto-de-Souza, Osmar 143

Paixão-Cortez, Vanessa Rodrigues 156
Pereira, Max 121
Peres, Liciana R.M.P. y 48
Ponzoni, Ignacio 36
Porto, William F. 147

Ramos, Daniela Fernandes 151
Rigo, Maurício Menegatti 160
Rizzi, Caroline 151
Roehe, Paulo Michel 156
Ruiz, Duncan D.D. 143
Rusu, Irena 24

Santos, Cássia T. dos 86
Seixas, Fabiana Kommling 151
Silva, Renato R.O. da 109
Silva, Tulio C. 73
Simão, Éder 164
Simões, Carlos 121
Sinigaglia, Marialva 164
Souto, Marcilio C.P. de 48
Spilki, Fernando Rosado 156

Tessier, Dominique 24
Togawa, Roberto C. 73

Vieira, Gustavo Fioravanti 160

Walter, Maria Emilia M.T. 73
Winck, Ana T. 143

Zelikovsky, Alex 1